Meyer—*Introduction to Mathematical Fluid Dynamics*
Montgomery and Zippin—*Topological Transformation Groups*
Morse—*Variational Analysis: Critical Extremals and Sturmian Extensions*
Nagata—*Local Rings*
Nayfeh—*Perturbation Methods*
Niven—*Diophantine Approximations*
Plemelj—*Problems in the Sense of Riemann and Klein*
Ribenbolm—*Algebraic Numbers*
Ribenbolm—*Rings and Modules*
Richtmyer and Morton—*Difference Methods for Initial-Value Problems,* 2nd Edition
Rivlin—*The Chebyshev Polynomials*
Rudin—*Fourier Analysis on Groups*
Samelson—*An Introduction to Linear Algebra*
Sansone—*Orthogonal Functions*
Siegel—*Topics in Complex Function Theory*
  Volume 1—Elliptic Functions and Uniformization Theory
  Volume 2—Automorphic Functions and Abelian Integrals
  Volume 3—Abelian Functions and Modular Functions of Several Variables
Stoker—*Differential Geometry*
Stoker—*Nonlinear Vibrations in Mechanical and Electrical Systems*
Stoker—*Water Waves*
Tricomi—*Integral Equations*
Wasow—*Asymptotic Expansions for Ordinary Differential Equations*
Whitham—*Linear and Nonlinear Waves*
Yosida—*Lectures on Differential and Integral Equations*
Zemanian—*Generalized Integral Transformations*

AN INTRODUCTION TO INVARIANT IMBEDDING

AN INTRODUCTION TO INVARIANT IMBEDDING

R. BELLMAN

G. M. WING

WILEY-INTERSCIENCE PUBLICATION

JOHN WILEY & SONS New York · London · Sydney · Toronto

Library of Congress Cataloging in Publication Data:
Bellman, Richard Ernest, 1920–
An introduction to invariant imbedding.

(Pure and applied mathematics)
"A Wiley-Interscience publication."
Includes bibliographical references.
1. Invariant imbedding. I. Wing, George Milton, 1923– joint author. II. Title.

QA431.B34 515 74-18455
ISBN 0-471-06416-5

Printed in the United States of America

10 9 8 7 6 5 4 3 2 1

PREFACE

The theory presented in this book is a relatively new one. Its major development, first as a method of studying some complex questions in astrophysics and more recently for investigating an extensive range of problems of scientific interest, has taken place largely over the last 25 years. As in all situations of this kind it is not difficult to find hints, glimmers, and precursors of the basic concepts 50 or 75 years earlier. Nevertheless, it was not until the astrophysicist S. Chandrasekhar laid the firm foundations of the method of invariance in his classic book *Radiative Transfer* [1] that a modern theory can be said to exist.

The authors became interested in transport phenomena while office mates at Los Alamos in 1945–1946. Our contact with such ideas as the point of regeneration method of the modern theory of stochastic processes and the ingenious use of functional equation techniques developed by R. Redheffer (see, e.g., [2,3]) led rather naturally to ultimate careful examination of Chandrasekhar's work on invariance. That, it is true, had been deeply influenced by earlier efforts of Ambarzumian involving the application of functional equation methods to the problem of radiative transfer in plane-parallel atmospheres. Chandrasekhar, however, was the first to develop a true theory from these concepts, a major contribution to modern astrophysics. Out of this background grew the theory and methodology that form the subject matter of this book. Although the authors are pleased to have played an early and a continuing role in this development, we hasten to point out that many other researchers have been involved and that their number is growing. We especially acknowledge the important contributions of one of our early collaborators, R. Kalaba.

The idea of *imbedding* a particular problem within a family of related problems is one of the fundamental conceptual devices of modern culture. It is practically the backbone of the comparative method which is used in such disparate fields as anatomy, religion, anthropology, and philology, as well as in classical mathematical physics, where the basic independent variables are space and time. Our investigations, of course, employ the imbedding device. The fact that both Chandrasekhar's concept of invariance and the all-pervasive idea of imbedding are joined has led to the term "invariant imbedding." From one viewpoint, invariant imbedding may be considered to be an application of perturbation theory. Thus, it seems to belong to an old and well-understood branch of analysis. However, the entity perturbed—namely, the structure of the system under study—is by no means a classical variable. It is this fact that makes the new theory novel and results in the need for both extensive and intensive investigations.

The increasing use of invariant imbedding as a practical and theoretical tool by scientists of many different persuasions seems to justify the preparation of a book on the subject at this time. In this way the general methodology will be made accessible to members of a relatively wide audience who may judge for themselves the relevance of the concepts involved to their own work. It is only through exposure of this kind that the invariant imbedding method will eventually find its valid place in the great panoply of applied mathematics.

It must be mentioned that discussions of invariant imbedding have appeared in other books. As noted above, it was first extensively used, under the name "Method of Invariance" by Chandrasekhar in his famous work on radiative transfer [1]. One of the current authors based about half of his treatment of transport theory on the method [4]. It has also been discussed over the past decade by R. Preisendorfer [5], R. Shimizu and R. Aoki [6], R. Bellman, R. Kalaba, and M. Prestrud [7], R. Bellman [8], K. Case and P. Zweifel [9], R. Adams and E. Denman [10], E. Denman [11], J. Mingle [12], and M. Ribarič [13]. However, only very recently have a few books appeared that have invariant imbedding as their primary focus. Hence this book is among the first efforts to present the method per se, with emphasis on the methodology itself rather than on the particular physical problems to which it can be applied.

This is not to say that we shall ignore applications of the technique. Our approach is to build up in the first three chapters a toolchest of invariant imbedding methods. We follow a somewhat historical presentation insofar as concepts are concerned, although our treatment avoids unpleasant and confusing complications often found in the original literature. This is accomplished basically within the context of a set of simple physical models, gradually leading to deeper mathematical insight into the general imbedding concept. The major part of the book consists of applications of the invariant imbedding method to specific areas, questions that are often of considerable direct interest to the engineer, physicist, applied mathematician, numerical analyst, and so on. A reader who masters the first three chapters can then select those particular subjects that specifically interest him in later chapters without having necessarily to read all of the intervening material.

Needless to say, the applications presented are primarily some of those that one or both of the authors have been involved in deeply. We make no pretense of giving a comprehensive coverage of all that is currently known about the subject of invariant imbedding. Indeed, there is relatively little overlap with the material in those treatises on the subject which have recently become available. This remark applies not only to the specific topics treated but also to the general approach and philosophy.

We shall avoid an "extreme" approach, which so often makes even books on applied mathematics essentially inaccessible to the very people who may wish to make use of the techniques described. Our reasons for this are the following: First—and this is by far the stronger—we feel that any such mathematical pseudo-sophistication is completely inappropriate for the bulk of the readers we hope to reach. Second, the level of development of the subject of invariant imbedding as a mathematical discipline is still neither extensive nor intensive. Although we shall not hesitate to state, and even to prove, a rigorous theorem when one is available or an elegant demonstration exists, we often indicate that some of the analysis is still at a rather heuristic stage. As a matter of fact, in a number of important cases no satisfactory mathematical results are yet available, despite the computational success of many of the procedures in numerous applications. This state of affairs may well provide incentives to those readers who are mathematically inclined to venture more deeply into the foundations of the subject.

As is often the case, applications of general methods of conceptual simplicity lead to analyses that can only be termed "messy." When the basic ideas are threatened with effective concealment by the technical details involved, we do not hesitate to sketch briefly the highlights of what is taking place and then refer the interested reader to the problems and to the original source material for further detail.

Although one of the prime values of invariant imbedding lies in its applications to calculations—especially massive computations which must be accomplished on high-speed digital computing machines—this part of the subject will not be emphasized. There already exists a rather extensive literature stressing such aspects of the subject, and elaborate tables of data of value to scientists in various fields have been printed. Hence, although numerical results will be presented here and there, they will be primarily in the form of examples. Furthermore, little will be said about the precise details of the finite difference schemes used, their convergence properties, stability, relative advantages and disadvantages, and so on. Indeed, many of these topics are in need of extensive further investigation and offer fruitful fields of research.

In order to allow the use of this volume as a textbook—or at the very least as a self-learning device for the reader who wants to become reasonably familiar with the new concepts involved—a large set of problems is included at the end of each chapter. Some of these are quite simple and straightforward; others are devised to cover some of the "messy" details previously mentioned. A still more complicated set is almost of the research variety, designed to bring the reader to the very frontier of current knowledge or even a bit beyond. Many of the problems are directly connected with invariant imbedding; others have only a tenuous relationship and are devised to acquaint the reader with types of equations or way of thinking that are uncovered by the technique. Thus numerous problems on such apparently disparate matters as Riccati equations, continued fractions, functional equations, and Laplace transforms are to be found. A "star" (asterisk) is used to indicate that a problem is difficult, lengthy, or of a nature peripheral to the basic subject matter of the text. Many of these starred problems are quite challenging. The reader whose primary goal is simply to gain a good understanding of the overall subject may not wish to attempt any of these. An ability to do the unstarred problems will assure the reader that he has an adequate grasp of basic matters.

Each chapter contains a list of references. This consists of articles and books specifically referred to in the text that are especially pertinent to the material of the chapter. Certain problems, usually of the starred variety, also contain references. We have avoided the temptation to try to present a vast bibliography on the overall subject of invariant imbedding. This is in part because, as noted earlier, no effort has been made to cover all aspects of the subject in this book. Moreover, several fairly complete bibliographies already exist. The early days of the subject may be traced through the references in [4]. In addition, two especially prepared bibliographies, one by T. Roşescu [14] and the other by M. Scott [15] are available. We urge the interested reader to consult these sources.

An effort has been made to use a relatively consistent notation in the first three chapters, since those are basic. The reader who reads further in the book may become somewhat irritated at such matters as x being always to the left of y in Chapter 10, while in Chapter 12 y is always to the left of x. Here the notation has been dictated in large part by the original research papers. We feel that anyone who wishes to delve deeply into particular subject areas will be less disturbed by notational variations from chapter to chapter than by digressions from the notation of the research literature. We have attempted consistency within any one particular chapter, however. Since the exercises are to present the reader with "real-life" situations, we make no effort to adopt any standard notation in the problem sets.

It seems appropriate to mention our style of presentation. At times the reader may feel that it suggests that the authors are personally responsible for everything described. It is true that each of us feels a very close involvement with the development of the subject. However, we have done our best, through specific references, to credit the many who have contributed. We trust that when we have failed to mention a particular author or a relevant research paper that it will be realized that we have erred unknowingly and we apologize for such omissions.

It is not possible to give proper credit to all those scientists who have in one way or another contributed to this book, through their own work, through conversations and letters, and, in many cases, through their remarks and comments at seminars and in the classroom. Nor can we begin to mention the many research agencies which over the last 15 years have so generously supported our research and writing efforts. We do wish to express hearty thanks to Mrs. Rebecca Karush who has typed the many versions of our manuscript and has courageously faced the problem of working with two authors who live a thousand miles apart. Finally, we express our appreciation to the staff of John Wiley and Sons, who have been so helpful in bringing our efforts to fruition.

R. BELLMAN
G. M. WING

Los Angeles and Los Alamos
June 1974

1. S. Chandrasekhar, *Radiative Transfer*, Dover, New York, 1960.
2. R. Redheffer, "Remarks on the Basis of Network Theory," *J. Math. Phys.* **28**, 1950, 237–258.
3. R. Redheffer, "Novel Uses of Functional Equations," *J. Rat. Mech. Anal.* **3**, 1954, 271–279.
4. G. M. Wing, *An Introduction to Transport Theory*, Wiley, New York, 1962.
5. R. W. Preisendorfer, *Radiative Transfer in Discrete Spaces*, Pergamon, Oxford, 1965.
6. R. Shimizu and R. Aoki, *Application of Invariant Imbedding to Reactor Physics*, Academic, New York, 1972.
7. R. Bellman, R. Kalaba, and M. Prestrud, *Invariant Imbedding and Radiative Transfer in Slabs of Finite Thickness*, American Elsevier, New York, 1963.
8. R. Bellman, *An Introduction to Matrix Analysis* (2nd ed.), McGraw-Hill, New York, 1970.
9. K. M. Case and P. Zweifel, *Linear Transport Problems*, Addison-Wesley, Reading, Mass., 1967.
10. R. Adams and E. Denman, *Wave Propagation and Turbulent Media*, American Elsevier, New York, 1966.

11. E. Denman, *Coupled Modes in Plasmas, Electric Media, and Parametric Amplifiers*, American Elsevier, New York, 1970.
12. J. O. Mingle, *The Invariant Imbedding Theory of Nuclear Transport*, American Elsevier, New York, 1973.
13. M. Ribarič, *Functional-Analytic Concepts and Structures of Neutron Transport Theory*, Slovene Academy of Sciences and Arts, Ljubljana, Jugoslavia, 1973.
14. T. Roşescu, "Invariant Imbedding in Neutron Transport Theory," Academia Republicii Socialiste România, Institutul' de Fizică Atomică, 1966.
15. M. Scott, "A Bibliography on Invariant Imbedding and Related Topics," Report SLA-74-0284, Sandia Laboratories, Albuquerque, N. M., 1974.

CONTENTS

Chapter 1 Fundamental Concepts 1

1 Introduction, 1
2 A Simple Physical Model, 2
3 A Classical Approach, 2
4 The Invariant Imbedding Approach, 6
5 Some Comments, Criticisms, and Questions, 9
6 A Minor Variant of the Model of Section 2, 9
7 A Major Variant of the Model of Section 2, 10
8 The Classical Approach Extended, 11
9 The Invariant Imbedding Approach Extended, 13
10 Some Comments on Possible Uses of the Reflection and Transmission Functions, 15
11 Summary, 16
Problems, 17
References, 20

Chapter 2 Additional Illustrations of the Invariant Imbedding Method 22

1 Introduction, 22
2 A Non-Linear Problem, 23
3 A Generalization of the Model, 24
4 Invariant Imbedding Formulation of the Model in Section 3, 25
5 The Linear Problem Revisited, 27
6 A Perturbation Approach, 28
7 Some Remarks and Comments, 31
8 The Riccati Transformation Method, 32
9 Summary, 35
Problems, 35
References, 38

Chapter 3 Functional Equations and Related Matters 39

1 Introduction, 39
2 A Basic Problem, 39
3 The Basic Functional Equations, 40
4 Some Applications of the Results of Section 3, 41
5 Differential Equations Via Functional Equations, 45
6 Summary, 47
Problems, 47
References, 53

Chapter 4 Existence, Uniqueness, and Conservation Relations 54

1 Introduction, 54
2 The "Physics" of the Conservative Case and Its Generalizations, 55
3 Another Derivation of the Reflection Function, 56
4 Some Conservation Relations, 57
5 Proof of Existence in the Conservative Case, 59
6 The Nonconservative Case: The Dissipation Function, 60
7 The Existence Proof, 63
8 Summary, 64
Problems, 65
References, 66

Chapter 5 Random Walk 67

1 Introduction, 67
2 A One-Dimensional Random Walk Process, 67
3 A Classical Formulation, 68
4 An Invariant Imbedding Formulation, 70
5 Some Remarks Concerning Section 4, 71
6 Sketch of Another Approach, 72
7 Expected Sojourn, 73
8 A "Many-State" Case—Invariant Imbedding Approach, 74
9 Time Dependent Processes—Classical Approach, 76
10 Time-dependent Processes—Invariant Imbedding Approach, 77
11 A Multistep Process—Classical Approach, 78
12 A Multistep Process—Invariant Imbedding Approach, 79
13 Some Remarks on an Extension to a Continuous Case, 81
14 Some Remarks About Random Walk in Two Dimensions, 82
15 Summary, 83

Problems, 83
References, 87

Chapter 6 Wave Propagation **88**

1 Introduction, 88
2 The Concept of a Plane Wave, 89
3 A Two Medium Problem, 89
4 A Multimedium Problem, 90
5 Resolution of the Multimedium Problem by "Wavelet Counting", 91
6 A Continuous Medium Problem, 94
7 An Analytical Approach to the Continuous Medium Problem, 97
8 The W.K.B. Method, 98
9 The Bremmer Series, 100
10 Another Imbedding, 103
11 Summary, 105
Problems, 105
References, 107

Chapter 7 Time-Dependent Problems **108**

1 Introduction, 108
2 A Time-Dependent Transport Problem—Particle-Counting Approach, 109
3 Time-Dependent Transport by Transform Techniques, 114
4 A Critique of the Foregoing, 116
5 Time-Dependent Input, 118
6 The Time-Dependent Wave Equation, 120
7 The Diffusion Equation, 121
8 Some Comments on the Previous Section, 124
9 A Critique of Sections 7 and 8, 125
10 Another Diffusion Problem, 126
11 A Final Diffusion Problem, 127
12 Summary, 129
Problems, 130
References, 132

Chapter 8 The Calculation of Eigenvalues for Sturm-Liouville Type Systems **133**

1 Introduction, 133
2 Eigenlengths for Transport-like Equations in One Dimension, 134

3 The Calculation of Eigenlengths, 134
4 Some Generalizations, 137
5 Results for Sturm-Liouville Systems, 138
6 Connection with the Prüfer Transformation, 140
7 Some Numerical Examples, 142
8 Summary, 144
Problems, 145
References, 146

Chapter 9 Schrödinger-Like Equations **147**

1 Introduction, 147
2 Formulation of the Phase Shift Problem, 148
3 A Representation of the Solution for Large t, 149
4 Partial Differential Equations for a and ψ, 151
5 Solution of the Partial Differential Equations for a and ψ, 153
6 Remarks on the Phase Shift Problem, 154
7 Formulation of the Eigenvalue Problem, 155
8 A Partial Differential Equation for $\tilde{b}$ and Its "Solution", 158
9 Resolution of the Difficulties, 161
10 Some Numerical Examples, 164
11 Some Remarks on the Eigenvalue Problem, 165
12 Summary, 166
Problems, 167
References, 168

Chapter 10 Applications to Equations with Periodic Coefficients **169**

1 Introduction, 169
2 Statement of the Problem, 170
3 The Differential Equations of Invariant Imbedding Over One Period, 171
4 Difference Equations Over an Integral Number of Periods, 171
5 Difference Equations Over a Nonintegral Number of Periods, 176
6 The "Backwards" Equations, 177
7 Some Numerical Results, 178
8 The Method of Doubling, 180
9 Trigonometry Revisited, 181
10 Summary, 182

Problems, 183
References, 185

Chapter 11 Transport Theory and Radiative Transfer 186

1 Introduction, 186
2 The Linearized Boltzmann Equation, 187
3 Some Remarks on Sections 1 and 2, 190
4 Boundary and Initial Conditions, 192
5 The Special Case of Slab Geometry and One Speed, 193
6 The Time-Independent Slab Problem via Invariant Imbedding—The Perturbation Approach, 196
7 The Time-Independent Slab Problem Via Invariant Imbedding—The Riccati Transformation, 202
8 A Return to the Case of the Semi-Infinite Half-Space, 206
9 Invariant Imbedding as a Calculational Device for Transport Problems in a Slab, 208
10 Transport Theory in Other Geometries, 209
11 Time-Dependent Transport in a Slab Geometry, 210
12 Summary, 214
Problems, 215
References, 217

Chapter 12 Integral Equations 219

1 Introduction, 219
2 An Integral Equation for Transport in a Slab, 220
3 A Pseudo-Transport Problem and Its Associated Integral Equation, 223
4 Representations for ϕ and n, 226
5 Derivation of the Principal Results, 227
6 A Special Case, 232
7 A Numerical Example and Some Remarks About Eigenvalues, 234
8 Further Remarks About the Foregoing, 238
9 A Completely Different Approach, 239
10 Summary, 242
Problems, 242
References, 245

Author Index 247

Subject Index 249

AN INTRODUCTION TO INVARIANT IMBEDDING

1

FUNDAMENTAL CONCEPTS

1. INTRODUCTION

In this chapter we shall investigate a very simple physical model which will lead quite readily to the basic idea of invariant imbedding. Actually, this model already occurs in a somewhat more specific form in the literature [1], arising naturally in a study of beta-ray transmission made by H. W. Schmidt in 1905. One may also find the imbedding concept used in a paper written some thirty-five years earlier by Stokes, who investigated the problem of light reflection from a stack of glass plates [2]. Although Stokes used the *number* of plates in the stack as a "structure" variable to arrive at a set of recurrence relations to resolve his problem, Schmidt continuously varied the thickness of his transmission medium and obtained corresponding differential equations. The overall functional equation approach has been used in the analysis of transmission lines and similar systems for many years (see, for example, Redheffer, [3,4]). Nevertheless, the continuous variation of system size does not seem to appear again in the literature until the classic work of Ambarzumian [5,6].

The problems investigated by this mathematically oriented astrophysicist were much more complex than any of those previously studied. However, it was not until the gifted mathematical physicist, S. Chandrasekhar, appeared and published his famous book on radiative transfer [7] that the authors and others began their intensive and extensive studies of the imbedding methods we shall describe in this book. Chandrasekhar developed an elegant theory of principles of invariance, thus completing and considerably extending the ingenious methods of Ambarzumian.

2. A SIMPLE PHYSICAL MODEL

We consider a "rod"—a line segment—which extends from 0 to x and denotes a generic point in the rod by z. "Particles" move to the right and left along this rod without interacting (that is, colliding) with one another. We shall, however, allow these particles to interact with the material of the rod itself. We assume at the moment that such interactions in no way affect the rod. They do, however, result in the disappearance of the particles involved in the collisions and the emergence of new particles, differing from the original only in that their direction of motion may have changed. For the time being, all moving particles are assumed to be of the same type and to have the same speeds. Our objective is to obtain information about the density of the beam of particles as a function of the position z.

Obviously, it is necessary to introduce more specific information if we are to have a mathematically meaningful problem. To this end we assume that the probability of a particle at z (moving in either direction) interacting with the rod while moving a distance $\Delta > 0$ is given by the expression

$$\sigma(z)\Delta + o(\Delta). \tag{1.1}$$

The notation $o(\Delta)$, which we use throughout this volume, signifies a term which goes to zero more rapidly than Δ as $\Delta \to 0$. The quantity σ is often called the macroscopic cross-section.

As noted earlier the immediate result of such an interaction is the disappearance of the particle. However, let us assume that as a consequence of a collision of this nature there emerge at the point z an average of $f(z)$ new particles going in the same direction as the original particle and an average of $b(z)$ new particles going in the opposite direction. Any right-moving particle emergent at $z = x$ is lost to the system, as is any left-moving particle emergent at $z = 0$. Finally, there are no spontaneous sources of particles in the rod. Particles injected at the left and right ends of the rod, together with progeny generated through the collision process, make up the total particle population of the system.

3. A CLASSICAL APPROACH

To simplify the initial analysis we shall suppose that a time independent state prevails; that is, the expected particle population is stationary, independent of the time at which the system is observed. Define

$$u(z) = \text{expected number of right-moving particles passing the point } z \text{ each second}, \tag{1.2}$$

$$v(z) = \text{expected number of left-moving particles passing the point } z \text{ each second}. \tag{1.3}$$

The word "expected" is clearly necessary in these definitions because of the fundamental stochastic nature of the process we have described. For convenience of expression, however, we shall often omit the word "expected" and simply ask that the reader bear it in mind.

We must now derive equations satisfied by the functions $u(z)$ and $v(z)$. To do so, let us examine a small portion of the rod lying between z and $z+\Delta$, and try to relate the function $u(z+\Delta)$ to the function $u(z)$. (We take $\Delta>0$ for convenience. See Figure 1.1.)

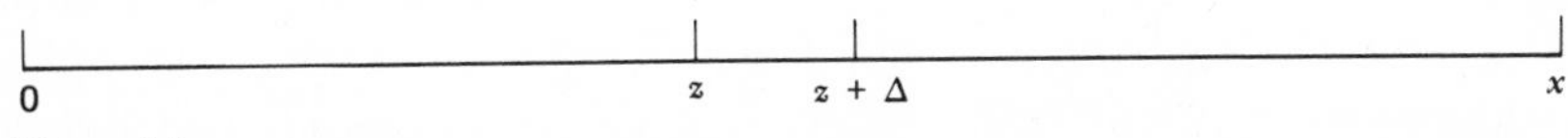

Figure 1.1

Certain of the particles passing the point z will move to the right as far as $z+\Delta$ without suffering any interactions with the rod. The probability of this happening is $(1-\sigma(z)\Delta)+o(\Delta)$, by assumption. Hence the expected number of such particles moving past z to the right contributing to $u(z+\Delta)$ is

$$[(1-\sigma(z)\Delta)+o(\Delta)]u(z)=(1-\sigma(z)\Delta)u(z)+o(\Delta), \tag{1.4}$$

provided all quantities involved are well defined and finite, an assumption that we shall ordinarily make.

We must also recognize that some of the right-moving particles passing z will interact with the rod before reaching the point $z+\Delta$. Each such event will produce an expected number $f(z)$ particles proceeding in the direction of interest. The expected contribution to $u(z+\Delta)$ from this type of occurrence is

$$[\sigma(z)\Delta+o(\Delta)]f(z)u(z)=\sigma(z)f(z)u(z)\Delta+o(\Delta). \tag{1.5}$$

Finally, we cannot completely neglect the particles that are moving to the left. Some of those which pass the point $z+\Delta$ will interact with the rod before reaching the point z and give rise to *right*-moving particles. The number of such particles then passing the point $z+\Delta$ each second is

$$[\sigma(z)\Delta+o(\Delta)]b(z)v(z+\Delta)=\sigma(z)b(z)v(z)\Delta+o(\Delta). \tag{1.6}$$

While one might at first suppose that many other events of significance can take place, a careful study of the model will reveal that all other contributions to $u(z+\Delta)$ are of order $o(\Delta)$. Hence

$$u(z+\Delta)=(1-\sigma(z)\Delta)u(z)+\sigma(z)\Delta f(z)u(z)+\sigma(z)\Delta b(z)v(z)+o(\Delta). \tag{1.7}$$

The reader who has followed this analysis with some care will note that a tacit assumption that all functions involved are continuous has been made. Equation (1.7) now yields

$$\frac{du}{dz} = \sigma(z)\{(f(z)-1)u(z)+b(z)v(z)\}. \tag{1.8}$$

If a similar "particle counting" process is carried out with attention focused on the left-moving particles the resulting equation is

$$-\frac{dv}{dz} = \sigma(z)\{b(z)u(z)+(f(z)-1)v(z)\}. \tag{1.9}$$

Equations (1.8) and (1.9) are the classical equations for this type of problem. They may be viewed as an extremely simple version of the fundamental Boltzmann equation of transport theory. We note, in particular, that attention has been confined to what is going on *within* the system, not on what is *emergent* from it. Indeed, the customary way of viewing most mathematical models of physical phenomena is analogous to the viewpoint we have employed. It is precisely this approach which is abandoned in the invariant imbedding method. However, before turning to a discussion of the new method, it is appropriate to continue the classical analysis that we have begun.

Our basic equations have been derived under the assumption that the functions u and v are time independent—that is, the system is in a stationary state. We must now impose boundary conditions compatible with that assumption. Let us suppose that each second one left-moving particle is injected into the rod at its right end, $z=x$, while no particles whatever enter the rod at the left end, $z=0$. Equations (1.8) and (1.9) must then be solved subject to the boundary conditions

$$u(0)=0, \qquad v(x)=1. \tag{1.10}$$

Now, it is mathematically well known that the system (1.8)–(1.10), a two-point boundary-value problem, may not have a solution. Even if it is soluble, it is possible that one or both of the functions u and v is negative somewhere. This is physically unreasonable in view of the definition of these quantities.

To understand better what may happen, let us carefully pick special functions f, b, and σ which ensure that an easy analytic solution to the problem is obtainable. We select

$$f(z)\equiv 1, \qquad b(z)\equiv 1, \qquad \sigma(z)\equiv 1, \tag{1.11}$$

so that the linear differential equations have constant coefficients. Then an elementary procedure gives

$$u(z)=\frac{\sin z}{\cos x},\qquad v(z)=\frac{\cos z}{\cos x}. \tag{1.12}$$

Observe that these solutions are mathematically meaningful provided that $x\neq(n+\frac{1}{2})\pi$. That is to say, the two point boundary value problem is soluble, with a unique solution, except for such x values. However, for *any* x value exceeding $\frac{1}{2}\pi$ the u and v functions become negative for some values of z in the interval $0\leqslant z\leqslant x$. Thus, the problem is not a physically sensible one unless $x<\frac{1}{2}\pi$.

The clue to this rather strange situation lies in the requirement that the system be time independent. Our particular choice of f and b implies that on the average two particles emerge from each collision interaction. It is therefore physically reasonable that the system may "go critical" if it is sufficiently large, that is, the u and v functions become unbounded. Physically, this is equivalent to saying that a stationary state cannot be maintained. Indeed, we have actually determined that the critical length of the rod is precisely $\frac{1}{2}\pi$ when the parameters are given by (1.11).

Using the explicit solution (1.12) we may very easily determine the number of particles emergent from the system. In view of the physics of the model, it is only natural to refer to the particles moving to the left at $z=0$ as "transmitted" and to those moving to the right at $z=x$ as "reflected." We define

$$\begin{aligned} t(x)&=\text{expected number of particles emergent per}\\ &\qquad\text{second at } z=0,\\ r(x)&=\text{expected number of particles emergent per}\\ &\qquad\text{second at } z=x, \end{aligned} \tag{1.13}$$

and refer to $t(x)$ and $r(x)$ as *transmission* and *reflection* functions, respectively. Observe that the argument of each function refers to the *length* of the rod and *not* to the point of emergence of the particles. Using the results obtained in (1.12) we find

$$t(x)=\sec x,\qquad r(x)=\tan x. \tag{1.14}$$

Note that when x approaches $\frac{1}{2}\pi$ from below each of these functions becomes unbounded. This is certainly consistent with any intuitive physical notion that one may have concerning the meaning of criticality.

An analytic solution for the functions u, v, t, and r is not ordinarily obtainable unless the functions f and b happen to be especially simple since linear second order differential equations with variable coefficients cannot usually be solved in terms of the elementary functions of analysis. Even the determination of the existence or nonexistence of a critical length is a very difficult problem in more general cases. In practical applications one must often turn to numerical techniques, which in view of the preceding remarks need not be routine.

4. THE INVARIANT IMBEDDING APPROACH

The method used in Section 3 provides us with complete information concerning the particle distribution in the rod. As a by-product we also obtain information concerning the reflection and transmission properties of the system. In some problems of genuine physical interest what happens *inside* the system is of little value; one is only interested in what emerges, the "observables." Such is the case in the problems already alluded to in Section 1 of this chapter. It is therefore reasonable to ask whether one can find equations satisfied directly by the r- and t-functions and thus by-pass completely the question of the internal particle distribution in the model. The method of invariant imbedding provides an affirmative answer to this question.

The initial reasoning is somewhat simplified if we assume the physical parameters, f, b, and σ to be constant over the entire rod length, $0 \leqslant z \leqslant x$. We now *imbed* this rod in one of slightly greater length, say $x+\Delta$, and having the same (constant) physical parameters as the original system. We define r and t precisely as in (1.13) and proceed to focus our attention on those physical events (collision interactions) which take place in the interval $x \leqslant z \leqslant x+\Delta$. Our goal is to find relationships connecting $r(x)$, $r(x+\Delta)$, $t(x)$, and $t(x+\Delta)$.

To begin, let us fix our attention on $r(x+\Delta)$, the expected number of particles emergent each second at the right end of the rod, assuming the same input as in the previous section. The physical events may be enumerated as follows:

a. The source particle has a collision upon passing through the initial interval $x \leqslant z \leqslant x+\Delta$. This gives rise to an average of b particles each second moving to the right and emergent at $x+\Delta$, thus contributing to $r(x+\Delta)$. It also produces an average of f particles moving to the left and hence impinging on the *sub*-rod that extends from 0 to x.

b. The source particle has no collision in the interval $x \leqslant z \leqslant x+\Delta$ and hence acts as a source particle for the *sub*-rod, $0 \leqslant z \leqslant x$.

The result of (a) and (b) is that the *sub*-rod now "sees" a certain input of particles; let us say temporarily that the strength of this source is s particles per second. The physical model we have constructed is such as to respond linearly to the input. Thus $sr(x)$ particles will, on the average, emerge each second at $z=x$. Again there are two possibilities:

a′. Some of these particles pass through $x \leqslant z \leqslant x+\Delta$ without interaction and so make a direct contribution to $r(x+\Delta)$.

b′. The remainder of these $sr(x)$ particles do experience collisions in passing through the interval $x \leqslant z \leqslant x+\Delta$. The outcome of each such event is an average of f particles moving to the right and hence contributing to $r(x+\Delta)$, together with b particles moving to the left and reentering the *sub*-rod $0 \leqslant z \leqslant x$. Such particles provide still another source for the *sub*-rod, and result in still more particles emergent at $z=x$.

One might now suppose that it is necessary to continue further, taking into account the collision interactions of the particles just discussed in the interval $x \leqslant z \leqslant x+\Delta$, and so on. Fortunately, such is not the case if Δ is a quantity we ultimately let approach zero. This we shall see in a moment by actually computing the number of particles involved in each of the above events.

Considering the events described in (a), we note that the probability of an interaction occurring in $x \leqslant z \leqslant x+\Delta$ is $\sigma\Delta + o(\Delta)$. Hence the expected number of particles "immediately" emergent at $x+\Delta$ is $\sigma b\Delta + o(\Delta)$, while an average of $\sigma f\Delta + o(\Delta)$ go on to the left. The probability of no such interaction taking place is $(1-\sigma\Delta)+o(\Delta)$. Therefore, event (b) contributes (each second) $(1-\sigma\Delta)+o(\Delta)$ particles to the source at $z=x$. Clearly we have $s = 1+(f-1)\sigma\Delta + o(\Delta)$.

Once again, using the probability of no interaction in $x \leqslant z \leqslant x+\Delta$, we find that (a′) contributes $sr(x)(1-\sigma\Delta)+o(\Delta)$ particles at $x+\Delta$. The right-moving particles from (b′) number $sr(x)\sigma f\Delta + o(\Delta)$ per second. The left-moving ones, averaging $sr(x)\sigma b\Delta + o(\Delta)$ per second, reenter the *sub*-rod at $z=x$. This is an effective source for the *sub*-rod, and it responds by yielding at $z=x$ a total of $\{sr(x)\sigma b\Delta + o(\Delta)\}r(x)$ more particles each second. These have probability of $(1-\sigma\Delta)+o(\Delta)$ of passing through the interval $x \leqslant z \leqslant x+\Delta$ without further interaction. The total contribution to $r(x+\Delta)$ enumerated thus far yields the relation

$$r(x+\Delta) = \sigma b\Delta + sr(x)(1-\sigma\Delta) + sr(x)\sigma f\Delta + sr^2(x)\sigma b\Delta + o(\Delta) \quad (1.15)$$

or

$$r(x+\Delta) = r(x) + \sigma\Delta\{b + 2(f-1)r(x) + br^2(x)\} + o(\Delta). \quad (1.16)$$

(Again, we have consistently assumed that all quantities involved are well defined and finite, and have handled the "o" terms accordingly).

Now if one attempts to calculate further contributions to $r(x+\Delta)$ he discovers that they are all of order $o(\Delta)$. This is because all further physical events must involve a total of at least *two* collisions in the interval $x \leqslant z \leqslant x+\Delta$.

Using (1.16) we obtain the following differential equation for r:

$$\frac{dr}{dx} = \sigma[b + 2(f-1)r(x) + br^2(x)]. \tag{1.17}$$

Similar reasoning (see Problem 8) leads to an equation for t:

$$\frac{dt(x)}{dx} = \sigma\{(f-1) + br(x)\}t(x). \tag{1.18}$$

It is important to note that Eq. (1.17) contains only the function r, while (1.18) involves both the r and t functions. This is a customary phenomenon associated with the imbedding method. [It is interesting to observe, however, that Schmidt, in his original work, obtained in a quite natural manner two equations, *each* of which involved both r and t (see Problem 3).]

Obviously, the equations we have derived are of little use without some side conditions which in this case are *initial conditions*. These are very easily found. Suppose the original rod is of length zero; thus $x=0$. In this case there can be no reflected particles, and all particles are transmitted. Therefore

$$r(0)=0, \qquad t(0)=1. \tag{1.19}$$

As a partial check on our reasoning, which is admittedly heuristic thus far, let us consider the special case given by Eq. (1.11). Simple calculations immediately lead to the solutions of (1.17)–(1.19):

$$r(x) = \tan x, \qquad t(x) = \sec x, \tag{1.20}$$

in complete agreement with (1.14).

The derivation we have given has been based upon the simplifying assumption that f, b, and σ are constant. (It is observed in Problem 4 that the assumption concerning σ is not an important one.) Let us now relax this condition and allow all three quantities to vary with position. Thus $f=f(z)$, and so on. We envision a rod of some fixed length $\bar{x}$. For $0 \leqslant z < \bar{x}$ the physical parameters of the rod are precisely the functions $f(z)$, $b(z)$, and $\sigma(z)$. We now consider the *sub*-rod extending from 0 to x and augment it by an additional length Δ. In this interval $x \leqslant z \leqslant x+\Delta$—the parameters

f, b, and σ are as prescribed. Obviously, this approach is valid so long as $x < \bar{x}$. If one now repeats the reasoning of this section Eq. (1.17) and (1.18) are obtained except that each time an f, b, or σ occurs it is evaluated at x. For example,

$$\frac{dr(x)}{dx} = \sigma(x)\{b(x) + 2(f(x) - 1)r(x) + b(x)r^2(x)\}. \tag{1.21}$$

The variable z never enters because all interactions which must be taken into account occur in the interval $x \leqslant z \leqslant x + \Delta$. Again, an assumption of continuity of the parameters is being made here. It is important to note that the continuity of σ, f and b may be considerably relaxed in certain cases (see Problem 5).

5. SOME COMMENTS, CRITICISMS, AND QUESTIONS

It has already been admitted that the approaches used in the preceding sections are heuristic. Both are, in the strict sense, imbedding techniques. However, the fundamental counting procedure in Section 4 is basically more complicated than that used in Section 3, or at least more unfamiliar. This makes the possibility for error seem great. The question then immediately arises: Isn't there an analytical way of obtaining the invariant imbedding results directly from the classical results—and conversely?

One might further ask if the method is confined to applications in relatively simple situations only or whether it can be extended to far more important and complicated models. Is it primarily a "physical" technique, or is it really a valid mathematical method that we have happened upon through physical reasoning? If the latter is the situation, perhaps the basic ideas, when mathematically exposed, can be applied to a wide variety of problems—perhaps even to those without obvious physical origins.

Suppose one had solved the r and t equations of Section 4 first. Is there a way of obtaining the u and v functions of the previous section making use of the knowledge of r and t without repeating all the work of Section 3? Are there any practical advantages in using the newer imbedding method instead of the more classical approach? Are the r and t equations numerically advantageous under certain conditions?

The above remarks and questions will be studied in the chapters to follow. Before turning to them, we introduce somewhat more complicated models.

6. A MINOR VARIANT OF THE MODEL OF SECTION 2

In our first simple model we assumed that an average of $f(z)$ particles emerged in the "forward" direction from each collision interaction, with

$b(z)$ particles moving in the "backward" direction. This is physically equivalent to saying that the rod does not distinguish between left- and right-moving particles when a collision occurs. Let us now drop this assumption and hypothesize that when a right-moving particle interacts with the rod the particle disappears and an average of $a_R(z)$ particles emerge at z moving to the right, while an average of $a_L(z)$ particles move to the left. However, when a left-moving particle is involved in such a collision interaction the respective expected numbers of particles produced are $b_L(z)$ to the left and $b_R(z)$ to the right.

Counting procedures similar to those used in Section 3 then yield

$$\frac{du}{dz} = \sigma(z)\{(a_R(z)-1)u(z)+b_R(z)v(z)\}, \tag{1.22a}$$

$$-\frac{dv}{dz} = \sigma(z)\{a_L(z)u(z)+(b_L(z)-1)v(z)\}. \tag{1.22b}$$

Although the details of the imbedding procedure are a bit more complicated now, the resulting equations are quite similar to those obtained in Section 4:

$$\frac{dr(x)}{dx} = \sigma(x)\{b_R(x)+[a_R(x)+b_L(x)-2]r(x)+a_L(x)r^2(x)\}, \tag{1.23a}$$

$$\frac{dt(x)}{dx} = \sigma(x)\{[b_L(x)-1]+r(x)a_L(x)\}t(x). \tag{1.23b}$$

We shall leave the derivations of these equations to the reader (see Problem 6). The various side conditions that result from the input specification are, of course, unchanged.

7. A MAJOR VARIANT OF THE MODEL OF SECTION 2

Suppose that the particles in the model of Section 2 are allowed to have different "states." Here we use the word "state" in a general sense—it might refer to the speed, energy, type, or any other feature (other than direction) that distinguishes among the particles. For simplicity, we shall assume a *finite* number, n, of such possible states, and denote them by indexing the corresponding functions. Let the probability that a particle in state j have an interaction collision with the rod material in passing from z to $z+\Delta$ be given by

$$\sigma_j(z)\Delta+o(\Delta). \tag{1.24}$$

Again we assume the disappearance of this particle and the emergence at z of new particles. However, we allow these new particles to have states different from the one involved in the original collision. Suppose an expected number $f_{ij}(z)$ particles come out traveling in the forward direction in the state i, while $b_{ij}(z)$ particles in state i emerge in the backward direction. We once more ask about the distribution of particles in the rod, and again suppose a time-independent state exists.

8. THE CLASSICAL APPROACH EXTENDED

To take into account the various possible particle states we define for $i=1, 2,\ldots,n$,

$$u_i(z) = \text{expected number of particles in state } i \text{ moving to the right past the point } z \text{ each second,} \tag{1.25}$$

$$v_i(z) = \text{expected number of particles in state } i \text{ moving to the left past the point } z \text{ each second.} \tag{1.26}$$

As in the earlier analysis, we attempt to relate the set of functions $\{u_i(z)\}$ to the corresponding set $\{u_i(z+\Delta)\}$. A counting procedure yields (for details, see, for example, [8]):

$$u_i(z+\Delta) - u_i(z) = \Delta\sigma_i(z)u_i(z) + \sum_{j=1}^{n} \Delta\sigma_j(z)f_{ij}(z)u_j(z) + \sum_{j=1}^{n} \Delta\sigma_j(z)b_{ij}(z)v_j(z+\Delta) + o(\Delta), \quad i=1,2,\ldots,n. \tag{1.27}$$

Similarly, if one focuses attention on the left-moving particles he obtains

$$v_i(z) - v_i(z+\Delta) = \Delta\sigma_i(z)v_i(z+\Delta) + \sum_{j=1}^{n} \Delta\sigma_j(z)b_{ij}(z)u_j(z) + \sum_{j=1}^{n} \Delta\sigma_j(z)f_{ij}(z)v_j(z+\Delta) + o(\Delta), \quad i=1,2,\ldots,n. \tag{1.28}$$

Obviously, if the usual continuity is assumed, Eq. (1.27) and (1.28) give rise to a system of differential equations, which are most easily expressed in matrix form. To this end, define the vectors,

$$u(z)=\begin{bmatrix} u_1(z) \\ u_2(z) \\ \vdots \\ u_n(z) \end{bmatrix}, \qquad v(z)=\begin{bmatrix} v_1(z) \\ v_2(z) \\ \vdots \\ v_n(z) \end{bmatrix}, \tag{1.29}$$

and the matrices,

$$F(z)=\begin{bmatrix} \sigma_1(z)(f_{11}(z)-1) & \sigma_2(z)f_{12}(z) & \cdots & \sigma_n(z)f_{1n}(z) \\ \sigma_1(z)f_{21}(z) & \sigma_2(z)(f_{22}(z)-1) & \cdots & \\ \cdots & \cdots & \cdots & \cdots \\ \sigma_1(z)f_{n1}(z) & & \cdots & \sigma_n(z)(f_{nn}(z)-1) \end{bmatrix}, \tag{1.30}$$

$$B(z)=\begin{bmatrix} \sigma_1(z)b_{11}(z) & \sigma_2(z)b_{12}(z) & \cdots & \sigma_n(z)b_{1n}(z) \\ \cdots & \cdots & \cdots & \cdots \\ \sigma_1(z)b_{n1}(z) & & \cdots & \sigma_n(z)b_{nn}(x) \end{bmatrix}. \tag{1.31}$$

The differential system may then be written

$$\begin{aligned} \frac{du}{dz} &= F(z)u+B(z)v, \\ -\frac{dv}{dz} &= B(z)u+F(z)v. \end{aligned} \tag{1.32}$$

Clearly, a wide variety of side conditions is possible. One set of particular interest to us is

$$u_i(0)=0, \qquad v_i(x)=\delta_{ij}.$$

(Here δ_{ij} is the Kronecker delta.) This is precisely the set which plays a large role in the invariant imbedding formulation to which we shall turn in the next section.

Finally, we remark without further comment that an extension of the model of Section 6 can be made in a similar way (see Problem 9).

9. THE INVARIANT IMBEDDING APPROACH EXTENDED

To carry out the imbedding procedure for the model now under consideration it is necessary to introduce a rather more complicated set of r and t functions. These are of great physical significance. We define

$r_{ij}(x)$ = the expected number of particles emergent each second at $z=x$ in state i from a rod of length x when the only input is one particle per second in state j at the right end, $z=x$,

$t_{ij}(x)$ = the expected number of particles emergent each second at $z=0$ in state i from a rod of length x when the only input is one particle per second in state j at the right end, $z=x$. (1.33)

Since $1 \leqslant i,\ j \leqslant n$, the functions r_{ij} and t_{ij} form the elements of square matrices. We shall often refer to these matrices as the reflection and transmission matrices and denote them simply by $R(x)$ and $T(x)$. They are occasionally referred to as *response* functions for the system described.

As one might suspect, the derivation of equations satisfied by $R(x)$ and $T(x)$ requires some patience using the particle-counting method. For some simplicity and variety let us obtain the result for $T(x)$, assuming that $R(x)$ is already known. Following the general ideas of the one-state case we consider a rod of length $x+\Delta$ and imbed in it a *sub*-rod of length x. We then attempt to relate $t_{ij}(x+\Delta)$ to $t_{ij}(x)$. In view of the form of Eq. (1.18) we should expect to find the elements of the R matrix entering in some way.

The basic physical events that must be accounted for in studying the transmission function $t_{ij}(x+\Delta)$ are as follows:

a. The source particle, in state j, has no collision in the initial interval $x \leqslant z \leqslant x+\Delta$. It therefore becomes a source particle in the same state for the *sub*-rod extending from zero to x.

b. The source particle has a collision in the initial interval $x \leqslant z \leqslant x+\Delta$. The backward-moving particles emerge at the right end $z=x+\Delta$ and make no contribution to the transmission. Those particles created in state j and moving to the left obviously contribute to the source in (a). Those particles created in state $k \neq j$ and moving to the left cannot, however, be neglected. More specifically, *all* the particles which the *sub*-rod "sees" entering in *any* state form a source for the *sub*-rod which can contribute *directly* to the transmission of particles emergent from $z=0$ in the required state i.

c. The *sub*-rod will *reflect* particles as a result of those impinging upon it at $z=x$. Only progeny of original source particles in state j that have had no collisions in $x \leqslant z \leqslant z+\Delta$ need be regarded. Such reflected particles may simply pass through the interval $x \leqslant z \leqslant x+\Delta$ and emerge at $x+\Delta$. In this case they can have no further effect since they are lost to the system. However, those reflected particles that *do* have interactions in the interval $x \leqslant z \leqslant x+\Delta$ can produce new "source" particles for the *sub*-rod, and these, regardless of their state, will contribute to $t_{ij}(x+\Delta)$.

The physical events enumerated above are the only ones that must be explicitly accounted for since all others involve at least two collisions in the interval $x \leqslant z \leqslant x+\Delta$. As usual, we shall assume all functions encountered are well defined and continuous. Hence, multiple collision events can contribute at most a term of order $o(\Delta)$.

Without further comment we write the expression for $t_{ij}(x+\Delta)$:

$$t_{ij}(x+\Delta)=(1-\sigma_j(x)\Delta)t_{ij}(x)+\sigma_j(x)\Delta\sum_{k=1}^{n} f_{kj}(x)t_{ik}(x)$$

$$+\sum_{p=1}^{n}\sum_{m=1}^{n} r_{mj}(x)\sigma_m(x)\Delta b_{pm}(x)t_{ip}(x)+o(\Delta). \qquad (1.34)$$

This leads to the system of differential equations

$$\frac{dt_{ij}}{dx}=-\sigma_j(x)t_{ij}(x)+\sigma_j(x)\sum_{k=1}^{n} f_{kj}(x)t_{ik}(x)$$

$$+\sum_{p=1}^{n}\sum_{m=1}^{n} r_{mj}(x)\sigma_m(x)b_{pm}(x)t_{ip}(x), \quad 1 \leqslant i,j \leqslant n, \qquad (1.35)$$

or, in more convenient matrix form,

$$\frac{dT(x)}{dx}=T(x)(F(x)+B(x)R(x)). \qquad (1.36)$$

The corresponding R equation, whose derivation we shall leave to the reader (Problem 8, see also [8]) is

$$\frac{dR(x)}{dx} = B(x) + R(x)F(x) + F(x)R(x) + R(x)B(x)R(x). \quad (1.37)$$

The side conditions are easily seen to be

$$R(0)=0, \qquad T(0)=I, \quad (1.38)$$

where I is the $n \times n$ identity matrix and the zero on the right-hand side of the first expression is the $n \times n$ zero matrix.

10. SOME COMMENTS ON POSSIBLE USES OF THE REFLECTION AND TRANSMISSION FUNCTIONS

In view of the apparent complexities of the equations just derived it is appropriate to remark on their possible utility. Of course, the same comments pertain as well to the simple case in which $n=1$.

While possibly rather forbidding at first sight, Eq. (1.36)–(1.38) are quite amenable to computation and to a good deal of rigorous analysis as well (see Chapter 4). Their principal advantage lies in the fact that they are initial value problems. Hence numerical integration is at least theoretically easier than is the numerical treatment of two-point boundary-value problems of the sort encountered in Section 8. (That this has actually proved to be the case in studies of physical importance will be pointed out more explicitly in future chapters.)

Moreover, knowledge of the reflection function can in itself be of value in studying the problem in its original classical formulation. To be specific, let us turn back to the two-point boundary-value problem posed by Eq. (1.8)–(1.10). Suppose the corresponding reflection problem (1.17) and (1.19) has been solved. Since $r(x)=u(x)$ is now known, the Eqs. (1.8) and (1.9) may be solved as an initial value problem subject to the two conditions at $z=x$,

$$u(x)=r(x), \qquad v(x)=1. \quad (1.39)$$

Again, in theory, this initial value problem is more easily solved than the original two-point problem. In practice, of course, basic questions of the stability of the integration method used enter into consideration, as they do in all formulations. Similar remarks can be made—and even more strongly—about the n-state case.

It should also be noticed that the linearity of the problems under consideration allows use of the superposition principle. Thus, for example, if the actual input at $z=x$ is s_p particles per second in state p and s_q per second in state q, then the corresponding number reflected each second in state i is $s_p r_{ip}(x)+s_q r_{iq}(x)$, and so on.

Change of the boundary conditions at the left end, $z=0$, is also possible. At the moment, the most obvious way of accomplishing such a change is to integrate the R system (1.37) subject to conditions other than $R(0)=0$. However, such a procedure destroys the superposition property that we have just mentioned. Since it will be seen later (see Chapter 3) that a much easier device is available, we shall pursue this matter no further here.

Finally, it is perfectly clear that there is nothing to prevent our considering a problem with zero input on the right and the source on the left. The details can be left to the reader (see Problem 10). In addition, there is the possibility of including internal extraneous sources of particles, that is, of particles that arise spontaneously internal to the system and are not the direct result of particle interaction with the rod. This phenomenon will be discussed in the chapters that follow.

11. SUMMARY

In this chapter we have introduced a very simple physical model of transport in a rod—a one-dimensional medium—and have analyzed the resulting problem in two different ways. The first method was the classical Boltzmann approach which depends upon concentrating on the distribution of particles within the system. The second and newer technique involved the introduction of the invariant imbedding method, in which primary emphasis is placed on the particles which actually *emerge* from the system. To some extent we have followed the historical development of the basic ideas, although even at this point it must be pointed out that the n-state rod case was not investigated by the imbedding device until after the much more complicated studies of Chandrasekhar were made on the transport problem in slab geometry (Chapter 11). The strict chronological approach is often not the best approach to a subject, no more that is a strict logical-mathematical foundation. It is often best to mix these in a flexible fashion.

We have mentioned some questions concerning the mathematical justification of our methods. It would seem, in fact, that what we have really obtained is a set of relations connecting linear differential equations with Riccati equations (*matrix* Riccati equations in the case of Section 9). Analogous relations have long been known in the theory of differential equations particularly as far as the second order differential equation is concerned. Now, however, we have endowed these relationships with certain physical meanings that are quite important. It begins to be apparent that the Riccati equation, named after an Italian count of the seventeenth century, is one of the fundamental equations of modern analysis and mathematical physics.

In the next two chapters we shall continue to pursue the study of relatively simple physical models in an effort to understand, both physically and mathematically, the fundamental concepts that we have introduced.

PROBLEMS

1. Discuss the possibility of reducing the pair of equations (1.8)–(1.9) to a single equation in either u or v in the case that b or f depends upon z. Try to find some explicit functions f and b for which this second order system can be solved.

2. In the text we have assumed as a standard hypothesis that f and b are continuous. If f and b are piecewise continuous, the Eqs. (1.8) and (1.9) hold on each interval of continuity. Physical considerations reveal that $u(z)$ and $v(z)$ must be continuous everywhere.

Consider a rod consisting of two contiguous materials such that $\sigma = 1$ and

$$f = 1, \qquad b = 1 \text{ for } 0 \leqslant z < 1;$$

$$f = \tfrac{1}{2}, \qquad b = \tfrac{1}{2} \text{ for } 1 \leqslant z \leqslant x.$$

Find $u(z)$ and $v(z)$ explicitly when $u(0) = 0$, $v(x) = 1$. Is the problem solvable for all x? For what values of x does the problem have a physically meaningful solution? Show that there is an x for which the system becomes "critical." Note that the material to the right of $z = 1$ thus results in a shorter critical length for the material to the left of $z = 1$ than was the case for the problem studied in the text (Section 3). Can you give a physical explanation for this phenomenon?

3. Consider the reflection and transmission problem for the case f, b, and σ constant. In the text the imbedding has been done at the end of the rod where the source is located. Try an imbedding at the other end, thus augmenting the rod at the left rather that the right. Obtain the r and t functions in this way. Observe that the resulting equations are completely coupled. This is the method used by H. W. Schmidt [1].

4. In the one-state case show that in both the classical and the invariant imbedding formulations the transformation

$$w = \int_0^z \sigma(z')\,dz'$$

leads to equations in which the "cross-section" σ no longer appears. Thus this transformation has the effect of replacing the variable σ by unity. In the astrophysical literature this is referred to as the "optical depth" transformation. Try to understand it physically. Is a similar device available for the n-state model?

5. In the invariant imbedding formulations the assumption has consistently been made that such functions as $f(z)$, $b(z)$, etc., are continuous. Relax this to piecewise continuity. Can you require even less?

6. Derive Eqs. (1.22) and (1.23).

7. In none of our discussions has the speed of the particle entered the discussions. In the n-state case suppose the particles in state j have speed c_j. Redo both the classical and the imbedding formulations to accommodate this, defining new functions when necessary.

8. Derive the R equation for the n-state case, and the t equation for the one-state case.

9. Extend the ideas of Section 6 to the n-state case.

10. Derive invariant imbedding equations for the various problems studied assuming the source is on the left instead of the right side of the system. Note that it is no longer satisfactory to fix the left end at $z=0$, and that the left extremity must now be a variable, say, $z=y$. Compare the equations you have obtained with those of the text. A certain "structure" should become apparent.

11. Suppose one keeps track of the "generation number" of the particles in the example of Section 3; that is, consider the incident particles at $z=x$ to be in the zeroth generation, all immediate progeny of these particles to be in the first generation, all immediate progeny of *these* to be in the second generation, and so on. It is now reasonable to define functions $u_n(z)$ and $v_n(z)$, where $u_n(z)$ is the expected number of nth generation particles passing z to the right each second, with a similar meaning for v_n. Derive classical "u and v" type equations. You will find that you have an infinite set of differential equations. Try to make the functions f and b also depend on the generation number. Note that each interaction results in an increase in the generation number of any particle of precisely one.

12. For the model described in Problem 11, develop corresponding r and t equations under the assumption that f and b do not depend on generation number. Thus $r_n(x)$ is the number of particles emerging each second from the right end of the rod with generation number n, and so on.

13. The generation number may actually be thought of as the *state* of the particle. The results obtained in the previous two problems should therefore be derivable directly from the discussion of the many state problem, even when f and b depend on generation number. Carry out this suggested program. (*Hint*. It is convenient in the invariant imbedding formulation to allow "fictitious" inputs of particles that are already in the nth generation for any n.) Note that while there can be an unbounded number of generations and the many state formulation has been studied only for a finite number of states, the text results may still be used directly. Explain why we can easily truncate.

14. In the light of Problem 13 and the generality of the idea of the "state" of a particle it would seem that one might discuss the generation number of a particle rather than its energy, for example. This is a valuable physical idea. Try to understand it, its advantages, and its shortcomings.

15. Suppose in a particular system particles are characterized by *both* their energies and their generation numbers. Derive both classical and invariant imbedding type equations, assuming only a finite number of possible energy states. (*Hint*. A well-thought-out notation will be extremely helpful!)

16. Let w satisfy the Riccati equation $w' = a(z) + b(z)w + c(z)w^2$. Show that $s = 1/w$ also satisfies a Riccati equation.

17. Show that if w is as in Problem 16 then $s = (\alpha w + \beta)/(\gamma w + \delta)$ likewise satisfies a Riccati equation. Consider the case in which α, β, and so on, are reasonably general functions of z.

18. Using the transformation in Problem 17 with appropriate choices of α, β, and so on, show how the Riccati equation in Problem 16 may be simplified in various ways. For example, how does one choose the coefficients to get an equation of the form $s' = d(z)s + e(z)s^2$? What choice yields $s' = f(z) + g(z)s^2$? Study other possibilities.

19. Suppose the coefficients in $w' = a(z) + b(z)w + c(z)w^2$ are analytic functions of z in some circle $|z| < r$. Under what conditions can the solution be written in the form $w(z) = \sum_{n=0}^{\infty} a_n z^n$?

20. Write the solution in Problem 19 as

$$w(z) = a_0 + a_k z^k h(z),$$

where $h(z)$ is given by a power series whose first term is not zero. Now set $h(z) = 1/s(z)$, so that $w(z) = a_0 + a_k z^k / s(z)$. According to the results of Problem 17 the function $s(z)$ also satisfies a Riccati equation. Find this Riccati equation explicitly.

Next note that $s(z)$ is analytic in some circle about the origin and so has a power series expansion

$$s(z) = \sum_{n=0}^{\infty} b_n z^n, \qquad b_0 \neq 0.$$

Now write $s(z) = b_0 + b_m z^m y(z)$ where $y(z)$ has a power series expansion whose first term is not zero. Set $y(z) = 1/q(z)$. Again $q(z)$ satisfies a Riccati equation that may be explicitly found.

If this process is continued one obtains a formal continued fraction expansion for the solution function w;

$$w(z) = a_0 + \cfrac{a_k z^k}{b_0 + \cfrac{b_m z^m}{c_0 + \cdots}}.$$

Obtain the first few terms of this expansion in terms of the coefficients of the power series for $a(z)$, $b(z)$, and $c(z)$.

This approach is due originally to Laguerre.

21.* In Problem 20 suppose the coefficients $a(z)$, $b(z)$, $c(z)$ also depend on a parameter ϵ in an analytic fashion and that the solution function $w = w(z, \epsilon)$ has an expansion $w = \sum_{n=0}^{\infty} a_n(z)\epsilon^n$. Use the ideas of Problem 20 to obtain a formal continued fraction expansion of w in terms of ϵ.

22. Consider the Riccati equation $w' = -w + w^2$, $w(0) = c$. Show that it is equivalent to the infinite system

$$w'_{n+1} = -(n+1)w_{n+1} + (n+1)w_{n+2}, \qquad w_n(0) = c^n, \qquad n = 0, 1, 2, \ldots.$$

(*Hint*. Try $w_n = w^n$.) This type of linearization is due to Carleman.

23.* Truncate the scheme in Problem 22:

$$\tilde{w}'_{n+1} = -(n+1)\tilde{w}_{n+1} + (n+1)\tilde{w}_{n+2}, \qquad n = 0, 1, 2, \ldots, N-2,$$

$$\tilde{w}'_N = -N\tilde{w}_N,$$

with $\tilde{w}_n(0) = c^n$, $n = 1, 2, \ldots, N$. Suppose that $|c| < 1$. Show that $\lim_{N\to\infty} \tilde{w}_n = w_n$.

24.* Let $p(z)$ be periodic with period 1. Under what further conditions on p, if any, does the equation $w' = p(z) - w - w^2$ possess a periodic solution of period 1?

25.* Consider the Riccati equation

$$w' + w + w^2 = ke^{-az}, \qquad w(0) = c.$$

Show that w has an expansion of the form

$$w = \sum_{n,m=0}^{\infty} w_{nm}(z)c^n k^m,$$

provided c and k are both sufficiently small. How can one determine the functions $w_{nm}(z)$ in an efficient manner?

26. Show that $w^2 = \max_y (2wy - y^2)$.

27.* The Riccati equation $w' = 1 - w^2$, $w(0) = c$, may be solved explicitly. It may also be written in the form [see Problem 26]

$$w' = 1 - \max_y (2wy - y^2) = \min_y (1 - 2wy + y^2).$$

This suggests that one examine the linear equation

$$s' = 1 - 2sy + y^2, \qquad s(0) = c,$$

where y is a quite arbitrary function. If the solution is denoted by $s(z,y)$ then it is reasonable to expect that $w(z) = \min_y s(z,y)$.

Verify this conjecture by explicitly solving the equation for $s(z,y)$. (See R. Bellman, *Methods of Nonlinear Analysis,* Vol. II, Chapter 11, Academic, New York, 1973.)

REFERENCES

1. H. W. Schmidt, "Reflexion u. Absorption von β Strahlen," *Ann. Physik*, **23**, 1907, 671–697.
2. G. C. Stokes, *On the Intensity of the Light Reflected or Transmitted through a Pile of Plates*, Mathematical and Physical Papers, **2**, University Press, Cambridge, 1883.

3. R. M. Redheffer, "Difference Equations and Functional Equations in Transmission Line Theory," in *Modern Mathematics for Engineers* (E. F. Beckenbach, Ed.), Second Series, Chapter 12, McGraw-Hill, New York, 1961.
4. R. M. Redheffer, "On the Relation of Transmission-line Theory to Scattering and Transfer," *J. Math. Phys.* **41**, 1962, 1–41.
5. V. A. Ambarzumian, "Diffuse Reflection of Light by a Foggy Medium," *C. R. Acad. Sci. SSR* **38**, 1943, 229.
6. V. A. Ambarzumian (Ambartsumian), *Theoretical Astrophysics*, Pergamon, New York, 1958.
7. S. Chandrasekhar, *Radiative Transfer*, Dover, New York, 1960.
8. G. M. Wing, *An Introduction to Transport Theory*, Wiley, New York, 1962.

2

ADDITIONAL ILLUSTRATIONS OF THE INVARIANT IMBEDDING METHOD

1. INTRODUCTION

In this chapter we shall gradually introduce several new approaches to the method of invariant imbedding. For the most part, we shall continue to examine simple physical models, although as we go on the reader will observe that we are more and more able to speak simply in terms of systems of differential equations, often neglecting their physical origins. Ultimately, we shall discover that the invariant imbedding method is quite independent of any basic physical considerations, and can be completely "mathematicized." At the same time, we shall very often find it convenient to continue to speak in physical terms, referring to "particles," "collisions," and the like. In fact, the importance of "particle counting" should not be ignored even when more sophisticated methods become known. These counting methods are of value, for example, in certain nuclear studies which use the essential imbedding concepts [1]. They also yield important analytic representation in equations connected with both particle and wave processes. Furthermore, the physical background often suggests powerful analytic and computational approaches, as well as methods of successive approximation producing upper and lower bounds.

The later chapters of this book will employ whichever imbedding approach seems clearest and most pertinent to the problem under consideration.

2. A NONLINEAR PROBLEM

Surprisingly enough, a considerable amount of insight into the fundamentals of invariant imbedding can be obtained by turning to the investigation of a nonlinear transport problem. Historically this model was first studied by the particle counting method of the preceding chapter [2]. The analysis proved so complicated and confusing that a simpler method was naturally sought, and eventually found. These new ideas were at once applied to the corresponding linear problems, and thus a new method of obtaining all the results of Chapter 1 became available.

Let us pose our nonlinear transport problem. Once again, we consider a rod (a line segment) extending from 0 to x, with z denoting a generic point on the rod. We define the probability of interaction of a particle with the rod material itself precisely as in the linear model. This is given, in the one-state case, by Eq. (1.1). Furthermore, $f(z)$ and $b(z)$ will have the same definitions as in Section 2 of Chapter 1. However, we now allow a new physical interaction. Particles moving to the right can *collide* with particles moving to the left. We assume that such a particle–particle interaction results in the annihilation of both particles—they simply disappear. To be more specific, we shall agree that if u particles pass z each second moving to the right and v particles pass $z+\Delta$ each second moving to the left then $\phi(u,v)\Delta+o(\Delta)$ particles are annihilated each second in the interval $(z,z+\Delta)$. For convenience, let us suppose that ϕ is a continuous and perfectly well behaved function of both u and v.

Before proceeding further, it is essential to observe that the u and v functions are now rather more complicated than was the case in the linear model. In particular, their values will depend in a significant way on the inputs at $z=0$ and $z=x$ as well as upon the position variable z, rather than in a "proportional" fashion as before. For convenience and consistency, we assume no input at $z=0$ and introduce y particles per second at $z=x$. We now define

$$u(z,y) = \text{the expected number of right-moving particles passing } z \text{ each second due to an input at the right end of the rod of } y \text{ particles each second,} \tag{2.1a}$$

$$v(z,y) = \text{the expected number of left-moving particles passing } z \text{ each second due to an input at the right end of the rod of } y \text{ particles each second.} \tag{2.1b}$$

We again consider a time-independent situation and hence make no effort to include the time variable in the definition of these functions. It is very easy to verify by "classical" particle counting methods that the equations satisfied by u and v are

$$\begin{aligned} \frac{\partial u}{\partial z} &= \sigma(z)\{(f(z)-1)u(z,y)+b(z)v(z,y)\} - \phi(u(z,y),v(z,y)), \\ -\frac{\partial v}{\partial z} &= \sigma(z)\{b(z)u(z,y)+(f(z)-1)v(z,y)\} - \phi(u(z,y),v(z,y)), \end{aligned} \tag{2.2}$$

with two-point boundary conditions

$$u(0,y)=0, \qquad v(x,y)=y, \qquad y \geqslant 0. \tag{2.3a}$$

Also

$$u(z,0)=v(z,0)=0, \qquad 0 \leqslant z \leqslant x. \tag{2.3b}$$

We pause to remark that the addition of the nonlinear term can have a most striking effect on the physical behavior of the model under discussion. For example, in the case $f=b=\sigma=1$, and $\phi(u,v)=\epsilon uv, \epsilon>0$, the system cannot go critical regardless of its length x. The interested reader is referred to [2, 3] for further results of this kind.

3. A GENERALIZATION OF THE MODEL

Instead of beginning an analysis via the imbedding method of the model just described we shall generalize even further. This tactic is not simply generalization for the sake of generalization. The form of (2.2) is sufficiently similar to that of (1.8) and (1.9) to suggest that an imbedding approach should essentially mimic that of the previous chapter. This was actually done when the above problem was originally encountered [2]. Deeper insight is achieved, however, if we write (2.2) in the general form

$$\begin{aligned} \frac{\partial u}{\partial z} &= f(u,v,z), \\ -\frac{\partial v}{\partial z} &= g(u,v,z). \end{aligned} \tag{2.4}$$

A finite difference version of (2.4) is

$$u(z+\Delta,y)=u(z,y)+f(u,v,z)\Delta+o(\Delta), \tag{2.5a}$$

$$v(z,y)=v(z+\Delta,y)+g(u,v,z)\Delta+o(\Delta). \tag{2.5b}$$

Let us consider (2.5) by itself for a moment without reference to its origin (2.2). We may interpret (2.5a) as stating that the number of particles moving each second to the right past $z+\Delta$ is the same as the number passing z to the right each second augmented (or diminished) by $f(u,v,z)\Delta$ particles, to within a quantity of order $o(\Delta)$. Similarly, (2.5b) states that the number of particles passing z to the left each second is the same as the number passing $z+\Delta$ to the left each second except for the augmentation (or diminution) of $g(u,v,z)\Delta$ particles, again to order $o(\Delta)$. Obviously, the specific form of the functions f and g is of no great consequence. In fact, had we merely encountered Eqs. (2.4) without being told of their physical origin, we could have given their finite difference versions this same interpretation. Thus, any system of the sort (2.4) can be abstractly viewed as arising from some transport problem. Naturally, questions of the solubility of the system under whatever side conditions may be imposed remain to be investigated in any particular case, and may be quite difficult.

4. INVARIANT IMBEDDING FORMULATION OF THE MODEL IN SECTION 3

We now turn to the problem of finding reflection and transmission functions for Eq. (2.4) subject to the side conditions given by (2.3). We define

$r(x,y)=$ the expected number of particles emergent at the right end of the rod, $z=x$, each second when the input is specified by (2.3), (2.6a)

$t(x,y)=$ the expected number of particles emergent at the left end of the rod, $z=0$, each second when the input is specified by (2.3). (2.6b)

If we wish at this point to speak somewhat more abstractly and avoid reference to "numbers of particles" we may simply note that

$$r(x,y)\equiv u(x,y), \qquad t(x,y)\equiv v(0,y), \tag{2.7}$$

We shall continue for the present to argue in terms of moving particles, however, and to think in terms of the physical model.

As in the previous chapter we augment the rod to length $x+\Delta$ and assume the original system to be a *sub*-rod imbedded in this new one. Let us begin by concentrating on r. From (2.5a)

$$r(x+\Delta,y)=u(x,y)+f(u,v,x)\Delta+o(\Delta). \tag{2.8}$$

There is now a strong temptation to replace $u(x,y)$ by $r(x,y)$. This, however, is incorrect. Since we are examining $r(x+\Delta,y)$, we are implying that while the rod has been changed in length by an amount Δ the input has remained y. The *sub*-rod of length x does *not* see the input y; it sees a slightly different input. Let us call this input temporarily $\tilde{y}$.

Clearly, we wish in (2.8) to remove the terms u and v which occur as arguments of f. These two functions must be replaced by their respective values at x, expressed in terms appropriate to the equation we are trying to develop, that is, in terms of r, t, and so on. Observe that u is just the number of particles passing $z=x$ to the right each second; in other words, the u argument of f is precisely r for the *sub*-rod. But that value is just $r(x,\tilde{y})$, where we have been careful to note that the second argument of this r function is the input $\tilde{y}$ which the system extending from zero to x sees. Similarly, the v argument of f is precisely the number of left-moving particles passing $z=x$ each second, and that is the input that the *sub*-rod sees, namely, $\tilde{y}$. We may now rewrite Eq. (2.8) as

$$r(x+\Delta,y)=r(x,\tilde{y})+f(r(x,\tilde{y}),\tilde{y},x)\Delta+o(\Delta). \tag{2.9}$$

The problem of finding $\tilde{y}$ remains to be resolved. As we have noted, $\tilde{y}$ is v at $z=x$ for the augmented system. Thus, from (2.5b):

$$\tilde{y}=y+g(r(x,\tilde{y}),\tilde{y},x)\Delta+o(\Delta), \tag{2.10}$$

where we have used the fact that $v(x+\Delta,y)$ is precisely the given input y, and we have replaced the u and v functions in g by their equivalents in the imbedding terminology. We may now rewrite (2.9):

$$r(x+\Delta,y)=r(x,y+g(r(x,\tilde{y}),\tilde{y},x)\Delta+o(\Delta))+f(r(x,\tilde{y}),\tilde{y},x)\Delta+o(\Delta). \tag{2.11}$$

Making our standard assumptions concerning continuity and the validity of differentiation, we obtain a partial differential equation for r:

$$\frac{\partial r}{\partial x}(x,y)-g(r(x,y),y,x)\frac{\partial r}{\partial y}=f(r(x,y),y,x). \tag{2.12}$$

The physics of the model provides the side condition

$$r(0,y)=0, \qquad y\geqslant 0. \tag{2.13}$$

Although, in general, (2.13) may not provide enough information for the complete solution of the partial differential system with which we are now confronted, it must suffice in this case if the problem is physically well-posed.

It is interesting to note that once again we have derived a "pure" r equation. Similar reasoning (Problem 2) provides an equation for t. As in the linear case, it is coupled to the equation for the reflection function:

$$\frac{\partial t}{\partial x}(x,y) = g(r(x,y)y, x)\frac{\partial t}{\partial y}(x,y). \tag{2.14}$$

The additional condition

$$t(0,y) = y, \qquad y \geqslant 0 \tag{2.15}$$

again follows readily from physical considerations.

5. THE LINEAR PROBLEM REVISITED

It is interesting to check the reasoning of Section 4 by applying the results to one of the linear problems examined in Chapter 1. Consider the problem posed in (1.22). Let us rewrite it in the form

$$\frac{d\tilde{u}}{dz} = a(z)\tilde{u}(z) + b(z)\tilde{v}(z), \tag{2.16a}$$

$$-\frac{d\tilde{v}}{dz} = c(z)\tilde{u}(z) + d(z)\tilde{v}(z). \tag{2.16b}$$

(We have chosen to write $\tilde{u}$ and $\tilde{v}$ instead of the u and v used in the previous chapter to avoid confusion with the somewhat more complicated functions introduced in Eq. (2.1). We have also simplified the coefficient notation somewhat, and trust that $b(z)$ will not be confused with the function used earlier.) In this case,

$$f = a(z)u(z,y) + b(z)v(z,y), \tag{2.17a}$$

$$g = c(z)u(z,y) + d(z)v(z,y). \tag{2.17b}$$

Moreover, because of the linearity of the problem, it is readily seen that

$$u(z,y) = \tilde{u}(z)y, \qquad v(z,y) = \tilde{v}(z)y, \tag{2.18a}$$

with

$$\tilde{u}(x) = 0, \qquad \tilde{v}(x) = 1. \tag{2.18b}$$

Furthermore, the same considerations yield

$$r(x,y) = \tilde{r}(x)y, \qquad t(x,y) = \tilde{t}(x)y. \tag{2.19}$$

(Again we have written the reflection and transmission functions depending only upon the one variable x with a tilde for notational consistency.)

Thus,

$$f=\{a(z)\tilde{u}(z)+b(z)\tilde{v}(z)\}y, \tag{2.20a}$$

$$g=\{c(z)\tilde{u}(z)+d(z)\tilde{v}(z)\}y. \tag{2.20b}$$

Substituting in (2.12) and (2.14) we obtain, after a bit of algebra,

$$\frac{d\tilde{r}}{dx}(x)=b(x)+\{a(x)+d(x)\}\tilde{r}(x)+c(x)\tilde{r}^2(x), \tag{2.21a}$$

$$\frac{d\tilde{t}}{dx}(x)=\{d(x)+c(x)\tilde{r}(x)\}\tilde{t}(x), \tag{2.21b}$$

in agreement with Eq. (1.23). The side conditions (2.13) and (2.15) readily lead to

$$\tilde{r}(0)=0, \qquad \tilde{t}(0)=1, \tag{2.22}$$

as usual.

We have thus confirmed that the method of the preceding section does indeed produce the same results as were obtained in the one-state case of the previous chapter. We note also that the technique used in the analysis of the nonlinear problem employed far less "particle counting" than was used in the study of the original linear problem, although it must be admitted that the derivations of (2.12) and (2.14) are still strongly based upon physical arguments.

6. A PERTURBATION APPROACH

A careful analysis of the invariant imbedding approaches used thus far suggests that some sort of perturbation method is actually concealed in our physical reasoning. The quantity being perturbed, however, is not a classical one. Rather it is the size—or length— of the system under study, which is to say the structure of the system. In most classical perturbation studies the size of the system remains fixed, and some other parameter is varied. In fact, it has jokingly been remarked that the secret of invariant imbedding lies in referring to the right end of the rod as x, classically considered a variable, rather than denoting it by a or b or some other letter that is customarily thought of as a constant!

With this idea in mind, let us try to formalize the perturbation technique and eliminate the physical reasoning from our derivations. To do this it is highly desirable to exhibit the parameter x explicitly in the definitions of our various functions. Thus we shall now write

$$u=u(z,x,y), \qquad v=v(z,x,y). \tag{2.23}$$

It will be convenient to denote partial differentiation by subscripting, using either the independent variable as the subscript or a numerical index subscript to denote the variable position. Hence,

$$\frac{\partial u}{\partial z}=u_z(z,x,y)=u_1(z,x,y), \qquad \frac{\partial v}{\partial x}=v_x(z,x,y)=v_2(z,x,y),$$

$$\frac{\partial r}{\partial y}=r_y(x,y)=r_2(x,y), \text{ etc.} \tag{2.24}$$

Let us rewrite Eq. (2.4) in this notation:

$$u_1(z,x,y)=f(u,v,z), \tag{2.25a}$$

$$-v_1(z,x,y)=g(u,v,z). \tag{2.25b}$$

We note that in all cases $0 \leqslant z \leqslant x$. The side conditions (2.3a) become

$$u(0,x,y)=0, \qquad v(x,x,y)=y. \tag{2.26}$$

Furthermore, the reflection and transmission functions are now

$$r(x,y)\equiv u(x,x,y), \qquad t(x,y)\equiv v(0,x,y). \tag{2.27}$$

We proceed to manipulate Eqs. (2.25) and (2.26) under our standard assumption that all differentiations and other operations are legitimate. First, we differentiate these equations with respect to the variable x:

$$u_{12}(z,x,y)=f_u(u,v,z)u_2(z,x,y)+f_v(u,v,z)v_2(z,x,y), \tag{2.28a}$$

$$-v_{12}(z,x,y)=g_u(u,v,z)u_2(z,x,y)+g_v(u,v,z)v_2(z,x,y), \tag{2.28b}$$

$$u_2(0,x,y)=0, \tag{2.28c}$$

$$v_2(x,x,y)=-v_1(x,x,y). \tag{2.28d}$$

Observe that under our assumptions

$$u_{12}(z,x,y)=u_{21}(z,x,y), \qquad v_{12}(z,x,y)=v_{21}(z,x,y).$$

If u and v are temporarily regarded as known functions, then Eqs. (2.28) are linear differential equations in the variable z for the unknown functions $u_2(z,x,y)$ and $v_2(z,x,y)$. Moreover, the boundary condition on the left, $z=0$, is the zero input condition to which we are accustomed.

Next, we differentiate (2.25) and (2.26) with respect to the variable y:

$$u_{31}(z,x,y)=f_u(u,v,z)u_3(z,x,y)+f_v(u,v,z)v_3(z,x,y), \quad (2.29a)$$

$$-v_{31}(z,x,y)=g_u(u,v,z)u_3(z,x,y)+g_v(u,v,z)v_3(z,x,y), \quad (2.29b)$$

$$u_3(0,x,y)=0, \quad (2.29c)$$

$$v_3(x,x,y)=1. \quad (2.29d)$$

Again regarding u and v for the moment as known we observe that this set of equations is of precisely the form we have been studying when regarded as a set of linear differential equations for the functions u_3 and v_3. The condition at $z=x$ is exactly the one we have imposed in Chapter 1.

Even more important is the observation that the systems (2.28) and (2.29) are identical in form except for the condition at $z=x$. If we assume that the functions f and g are such that a unique solution to (2.29) exists then the solution to (2.28) can be written down at once because of the linearity:

$$u_2(z,x,y)=-v_1(x,x,y)u_3(z,x,y), \quad (2.30a)$$

$$v_2(z,x,y)=-v_1(x,x,y)v_3(z,x,y). \quad (2.30b)$$

We also recognize that $-v_1$ can be replaced by g by virtue of Eq. (2.25b).

To obtain an expression for r we must obviously evaluate all functions involved at $z=x$. From (2.30a)

$$u_2(x,x,y)=g(u,v,x)u_3(x,x,y). \quad (2.31)$$

But from (2.27)

$$r_x(x,y)=u_1(x,x,y)+u_2(x,x,y), \quad (2.32)$$

while (2.25a) yields

$$u_1(x,x,y)=f(u,v,x). \quad (2.33)$$

Again using (2.27)

$$r_y(x,y)=u_3(x,x,y). \quad (2.34)$$

Combining these various results, noting the value of v at $z=x$ as given by (2.27), and using a consistent notation, we finally obtain the equation

$$r_x(x,y)-g(r(x,y),y,x)r_y(x,y)=f(r(x,y),y,x) \quad (2.35)$$

in complete agreement with (2.12). A similar argument based upon (2.30b) produces the t equation (2.14) (see Problem 3).

For the side conditions we turn to (2.26) and (2.27):

$$r(0,y)=u(0,0,y)=0, \tag{2.36a}$$

$$t(0,y)=v(0,0,y)=y. \tag{2.36b}$$

Actually, Eqs. (2.30) contain a good deal of information which we have not utilized, since immediately upon obtaining them we chose to set $z=x$. Historically, their value was not recognized at once, and they went neglected for some time. To discuss their usefulness here would unfortunately occasion too great a digression.

7. SOME REMARKS AND COMMENTS

Thus far in this chapter we have attempted to clarify the basic invariant imbedding method. We now recognize that it is indeed closely related to relatively familiar perturbation theory. No effort has as yet been made here to put the technique on a rigorous foundation, that is, to place conditions on f, g, x, y, and so on that make the reasoning of the preceding section mathematically impeccable. This could now be done at the expense of certain restrictions, such as the magnitude of x, but we shall not go through the details. To do so would divert us from our present course, which is to present the basic philosophy involved to the reader who is primarily interested in applications. Moreover, we should very likely find that the conditions we might impose in the perturbation analysis are not actually satisfied in cases of actual application. In later chapters during the discussion of some important problems we shall take a more rigorous viewpoint and be much more specific in our hypotheses.

We have not mentioned at all in the present analysis the n-state case which occupied a fair amount of our time in Chapter 1. The generalizations of Sections 4 and 6 to this case are quite straight forward (although rather cumbersome), and we prefer to leave the details to the reader (see Problem 5). Although it is too early to enter into any great detail, it should be mentioned that the perturbation method is also applicable to problems considerably more complex than the n-state model. For example, it can be applied to the problems of original interest to Ambarzumian and Chandrasekhar, which involve partial differential integral equations. This, incidentally, is an example of a situation of great practical interest which would *not* be mathematically clarified by a rigorous investigation of the model of this chapter. There is simply too wide a gap between our simple model and the Boltzmann equation of vital importance to the astrophysicist.

Another comment concerning the form of Eqs. (2.12) and (2.14) is pertinent. These partial differential equations are of a fairly familiar type. Suppose that we were confronted by these equations without any knowledge of how they arose. A standard method of attack is to write down the equations for their characteristic curves. For (2.12) these are easily seen to be

$$\frac{dx}{ds} = 1, \tag{2.37a}$$

$$\frac{dy}{ds} = -g(r,y,x), \tag{2.37b}$$

$$\frac{dr}{ds} = f(r,y,x). \tag{2.37c}$$

But these, apart from notation, are equivalent to Eqs. (2.4). That is, the equations with which we started are exactly the characteristic equations for the invariant imbedding equations. Have we then merely rediscovered a theory that has been understood and studied for well over a century? In one sense the answer to this question is "Yes!" In fact, even when the imbedding ideas are applied to such complicated cases as the Boltzmann equation alluded to in the previous paragraph, the results obtained may be considered to be generalizations of the classical theory of first order partial differential equations (see [4, 6]). However, this approach is so complicated and abstract that it is not at all easy to apply, and the authors venture to guess that had the more "homey" methods described in this book not come into being—in other words, had only the abstract characteristic theory been known—the imbedding method would be used in applications far less than it is today. We feel that it is fortunate that the theory is now understood at a variety of different levels of physical and mathematical sophistication.

8. THE RICCATI TRANSFORMATION METHOD

We wish to conclude this chapter by presenting still another approach to imbedding. It has already been noted in Chapter 1 that the equation for the reflection function in the n-state linear model is a generalized Riccati equation. Strong use of this fact has been made by G. Rybicki and some of his coworkers [5]. Although our presentation of their technique will seem somewhat unmotivated at the moment, the basic idea will become much clearer when we discuss functional equations in Chapter 3. Since, however, the Riccati transformation is another tool by which imbedding equations can be obtained, it seems appropriate to include it in this chapter rather than to delay its introduction.

Let us once again consider the one-state rod problem as described in Section 2 of Chapter 1. For variety and further insight, however, we now assume that the rod contains (spontaneous) sources of particles. Specifically, let us suppose that each second there appear in the interval, $(z, z+\Delta)$, an expected total of $s^{+}(z)\Delta + o(\Delta)$ particles moving to the right, and an expected total of $s^{-}(z)\Delta + o(\Delta)$ to the left. Equations (1.8) and (1.9) must now be replaced by

$$\frac{du}{dz} = \sigma(z)\{(f(z)-1)u + b(z)v\} + s^{+}(z), \tag{2.38a}$$

$$-\frac{dv}{dz} = \sigma(z)\{b(z)u + (f(z)-1)v\} + s^{-}(z). \tag{2.38b}$$

We shall assume the inputs at $z=0$ and $z=x$ are as in the original model:

$$u(0)=0, \qquad v(x)=1. \tag{2.39}$$

Admittedly without much motivation at the present time, we define functions $\rho(z)$ and $w(z)$ by the relation

$$u(z) = \rho(z)v(z) + w(z). \tag{2.40}$$

(Observe that we are reverting to the single independent variable notation of Chapter 1.) Differentiating (2.40) we obtain

$$\frac{du}{dz} = \rho(z)\frac{dv}{dz} + \frac{d\rho}{dz}v(z) + \frac{dw}{dz}. \tag{2.41}$$

By using (2.38) we obtain, after some algebra and rearrangement,

$$v(z)\left\{\sigma(z)b(z) + 2\sigma(z)(f(z)-1)\rho(z) + \sigma(z)b(z)\rho^2(z) - \frac{d\rho}{dz}\right\}$$
$$= \frac{dw}{dz} - \sigma(z)\{(f(z)-1) + b(z)\rho(z)\}w(z) - \rho(z)s^{-}(z) - s^{+}(z). \tag{2.42}$$

As yet, the functions $\rho(z)$ and $w(z)$ are not completely specified. We now specify them by requiring that they satisfy the following two equations

$$\frac{d\rho(z)}{dz} = \sigma(z)\{b(z) + 2(f(z)-1)\rho(z) + b(z)\rho^2(z)\}, \tag{2.43}$$

$$\frac{dw(z)}{dz} = \sigma(z)\{(f(z)-1) + b(z)\rho(z)\}w(z) + \rho(z)s^{-}(z) + s^{+}(z), \tag{2.44}$$

with the initial conditions

$$\rho(0)=0, \qquad w(0)=0. \tag{2.45}$$

Let us suppose that this system is soluble on the interval $0 \leqslant z \leqslant x$. [Note that this is basically a requirement on the equation for $\rho(z)$. The linearity of (2.44), together with our standard assumption that $\sigma(z)$, $f(z)$, $s^+(z)$, and so on , are well-behaved then guarantees that $w(z)$ exists.] Equation (2.42) is then satisfied for $0 \leqslant z \leqslant x$. Since $v(x)=1$, the value of $u(x)$ can be calculated from (2.40). Thus we have found the number of particles emergent from the right end of the rod. [Observe that (2.45) assures that $u(0)=0$.]

It is a matter of considerable interest that Eq. (2.43) is precisely the equation for r which was derived in Section 4 of Chapter 1. Since the initial conditions are the same for the r and ρ equations, it follows that $\rho(x)\equiv r(x)$. However, the function ρ has mathematical meaning—and likely physical meaning as well—at all points z of the rod, $0 \leqslant z \leqslant x$. On the other hand, r has until now been regarded as representing the number of particles *emergent* from a (source-free) rod. The present approach has thus apparently given r a meaning even *internal* to the system under study. All of this will become much clearer in the chapter which follows, and we shall obtain much insight into this Riccati transformation approach at that time.

That in general $u(x)\neq r(x)$ in this problem is not surprising, since we have introduced the extraneous source terms $s^+(z)$ and $s^-(z)$. It is easily verified that if these two terms are zero, $0 \leqslant z \leqslant x$, then indeed $w(z)\equiv 0$, and $r(x)$ is precisely the number of particles reflected from the right end of the rod. Thus our results are completely consistent with those of Chapter 1.

Our analysis may be profitably carried a bit further. From (2.38b) and (2.40) we readily obtain

$$-\frac{dv}{dz}=\sigma(z)\{b(z)\rho(z)+(f(z)-1)\}v(z)+\sigma(z)b(z)w(z)+s^-(z). \tag{2.46}$$

This equation is linear in $v(z)$ and may be integrated backwards from $z=x$ to $z=0$, starting with the condition $v(x)=1$. Since both $\rho(z)$ and $w(z)$ are already known, u may be obtained at any internal point z from the simple algebraic relation (2.40).

In the scheme that we have just described the function $\rho(z)$, which we have noted is identical to r, plays the key role. It satisfies a nonlinear equation; all other functions of interest can be obtained from equations of a much simpler form involving this function. Thus, the function which we have come to consider as the reflection function for the system seems again to be of primary mathematical and even physical importance in the analysis of transport-like problems.

It is easy to see that the arguments we have employed can be generalized without extensive modification to systems of the form

$$\frac{du}{dz}=A(z)u+B(z)v+s^{+}(z),$$

$$-\frac{dv}{dz}=C(z)u+D(z)v+s^{-}(z), \tag{2.47}$$

$$u(0)=0, \qquad v(x)=c,$$

where $u(z)$, $v(z)$, and $s^{+}(z)$, $s^{-}(z)$ are column vectors with all upper case expressions denoting square matrices. Once again, the generalized Riccati equation for the "reflection" function plays the essential role. Details are left to the reader (see Problem 7).

There does not seem to be an obvious way to extend this approach to the nonlinear problems studied in the earlier section of this chapter. Until quite recently relatively little work of a practical or computational nature has been done with invariant imbedding as applied to nonlinear problems (for an interesting treatment see [6]). Such equations were introduced primarily to give us further insight into the imbedding method rather than because of their own intrinsic importance to us.

9. SUMMARY

In this chapter we have presented several additional approaches to the method of invariant imbedding. Except for the last one, these techniques are applicable to nonlinear problems. Indeed, the study of such problems has provided deeper understanding of the method and its relationship to more classical techniques of analysis. Furthermore, we have progressed to a point where ways of making at least some of the methods rigorous are now reasonably clear. Nonetheless, we have chosen not to pursue this question very far. In the chapter that follows we shall take up the general subject of functional equations which arise naturally from the imbedding technique. These will, in fact, provide still further means of obtaining some of the fundamental differential equations for reflection and transmission functions.

PROBLEMS

1. Prove that when $f=b=1$ and $\phi(u,v)=u(z)v(z)$ the model represented by Eq. (2.2) cannot become critical. Try to generalize this to the case of other functions and also to other values of f and b, still assuming $f+b>1$.

2. Derive Eqs. (2.14) and (2.15), using the methods of Section 4.

3. Derive Eqs. (2.14) and (2.15) using the perturbation approach.

4. Formulate some reasonable conditions on the functions $f(u,v,z)$ $g(u,v,z)$ and the magnitude of x and y to make the perturbation approach of Section 6 rigorous.

5. Extend the results of Sections 1–7 to the n-state case.

6. Derive Eqs. (2.38) from physical principles. We have implicitly assumed continuity of the functions s^+ and s^-. Can this be relaxed?

7. Using matrix notation and methods throughout, extend the "Riccati" method to Eqs. (2.47), obtaining equations for both the reflection and transmission functions in the homogeneous case as well as for the solution functions u and v in the nonhomogeneous problem stated.

8. In the matrix case our analysis has always implied that there are as many u equations as v equations, and hence that there are as many conditions at $z=x$ as at $z=0$. Suppose there are more u equations than v equations. Introduce additional functions $v_i(z)$ and additional equations of the form $-v_i'(z)=\cdots$ so as to retain the solution of the given problem and in such a fashion as to allow the use of the invariant imbedding methods we have been discussing. Study the extraneous r_{ij} and t_{ij} functions that have thus been introduced and try to understand them.

Repeat the analysis for the case in which there are fewer u equations than v equations.

9. Consider the Riccati transformation in the source-free case. As noted in the text it reduces to $u(z)=r(z)v(z)$. Try to understand the physical meaning of this. Also observe that if $v(z_1)=0$ then $u(z_1)=0$ unless $r(z_1)$ fails to exist. The differential equations being considered are of such a nature that the usual fundamental existence and uniqueness theorems apply. Therefore, u and v cannot vanish at the same point unless u and v are identically zero. Try to interpret these observations both physically and mathematically.

10. Turn to Eqs. (2.47) and suppose $s^+=s^-=0$. Replace this system by the system

$$\frac{dU(z)}{dz}=A(z)U(z)+B(z)V(z),$$

$$-\frac{dV(z)}{dz}=C(z)U(z)+D(z)V(z),$$

where U and V are square matrices. Let $U_1(z)$ and $V_1(z)$ be a fundamental matrix solution. Define $\tilde{R}(z)=U_1(z)V_1^{-1}(z)$. Show that $\tilde{R}$ satisfies the standard reflection equation. Observe that the formula for the derivative of $V_1^{-1}(z)$ is the key to the Riccati equation. (For further details see Section 3 of Chapter 4.)

11. Consider an object subject only to the force of gravity and air resistance. Let it be propelled vertically with an initial velocity v_0. The equation defining its height y above the ground at any time t is

$$y''(t)=-g-h(y'), \qquad y(0)=0, \qquad y'(0)=v_0.$$

Assume that h is a nonnegative function and $v_0>0$. Show that its maximum altitude is given by

$$\int_0^{v_0} \frac{s\,ds}{g+h(s)}.$$

12. The problem given above is a classical one. Its interest here lies in the fact that it may also be solved by an imbedding technique. Let $f(v_0)$ be the maximum altitude achieved above the launch position. Show that $f(v_0)=v_0\Delta+f(v_0-(g+h(v_0))\Delta)+o(\Delta)$. Hence derive the foregoing result.

13. Generalize Problems 11 and 12 by assuming $y''(t)=k(y,y')$, with $y(0)=c_1\geqslant 0$, $y'(0)=c_2\geqslant 0$, and k a nonpositive function. Let $f(c_1,c_2)$ be the maximum altitude attained above the zero position. Prove that $c_2\partial f/\partial c_1+k(c_1,c_2)\partial f/\partial c_2=0$, $f(c_1,0)=c_1$. Show that the results of the previous problems are contained in this answer. Discuss possible computational use of this more general equation. (See R. Bellman, "Functional Equations and Maximum Range," *Quart. Appl. Math.* **17**, 1959, 316–318.)

14.* The scalar equation $dy/dt=g(y), y(0)=c$ may be approximated by the simple difference equation $y(t+\Delta)=y(t)+g(y(t))\Delta, y(0)=c, t=0,\Delta,\ldots$ for both analytical and computational purposes. Another approximation is

$$y(t+\Delta/g(y))=y(t)+\Delta, \qquad t=0,\Delta,\ldots.$$

Obviously two different imbeddings are involved. Discuss these, their meanings, and the possible advantages and disadvantages of each.

15.* The equation $dy/dt=\sqrt{y}, y(0)=0$, has the solution $y(t)=0$ and also the solution $y(t)=t^2/4$. Provide a finite difference approximation that will produce the second solution.

16. The partial differential equation $w_t=ww_x, w(x,0)=g(x)$ is satisfied by $w(x,t)=g(x+w(x,t)t)$. This expression defines $w(x,t)$ implicitly. Suppose $g(x)=x$. In this case, $w(x,t)=x/(1-t)$. Thus w becomes infinite as t approaches unity. This is a classical simple model of a hydrodynamical shock wave, which is somewhat analogous to the critical length phenomenon of transport theory. Show that the choice $g(x)=x$ is by no means necessary and that this shock phenomenon occurs for a very wide class of initial functions g.

17.* Show that the equation in Problem 16 is solved approximately by $w(x,t+\Delta)=w(x+w(x,t)\Delta,t)$. How can this equation be used to obtain numerical results? Note that the difference equation can be used to actually integrate through shocks. Is it useful in this region? (See R. Bellman, I. Cherry, and G. M. Wing, "A Note on the Numerical Integration of a Nonlinear Hyperbolic Equation," *Quart. Appl. Math.* **16**, 1958, 181–183. For a discussion of the convergence properties of the difference equation as $\Delta\to 0$, see R. Bellman and K. Cooke, "Existence and Uniqueness Theorems in Invariant Imbedding. II: Convergence of a New Difference Algorithm," *J. Math. Anal. Appl.* **12**, 1965, 247–253.)

REFERENCES

1. J. O. Mingle, *The Invariant Imbedding Theory of Nuclear Transport*, American Elsevier, New York, 1973.
2. R. Bellman, R. Kalaba, and G. M. Wing, "Invariant Imbedding and Neutron Transport. III: Neutron–Neutron Collision Processes," *J. Math. Mech.* **8**, 1959, 249–262.
3. T. A. Brown, "The Existence and Uniqueness of the Solution to a Problem in Invariant Imbedding," *J. Math. Appl.* **11**, 1965, 236.
4. G. H. F. Meyer, *On a General Theory of Characteristics and the Method of Invariant Imbedding*, Computer Science Center, Univ. of Maryland, Tech. Rep. TR-66-37, Dec. 1966.
5. G. B. Rybicki and P. D. Usher, "The Generalized Riccati Transformation as a Simple Alternative to Invariant Imbedding," *Astrophys. J.* **146**, 1966, 871–879.
6. G. H. Meyer, *Initial Value Methods for Boundary Value Problems*, Academic, New York, 1973.

3

FUNCTIONAL EQUATIONS AND RELATED MATTERS

1. INTRODUCTION

In this chapter we shall primarily study functional equations that relate reflection and transmission functions to one another and to the more classical "u" and "v" functions. These results are of great importance in themselves; furthermore, they provide additional means of deriving many of the imbedding equations we have already found by other devices. By the end of this chapter we shall have enough information concerning the basic concepts and tools of invariant imbedding to allow us to turn to a variety of problems of genuine physical interest.

Our attention in this chapter will be confined entirely to linear problems. For convenience in presentation we shall study the one-state case, although many of our results generalize readily to the many-state systems. These generalizations are left to the reader in the form of problems. Although we shall frequently continue to use physical terminology in our discussions, the astute reader will note that this is almost solely for the purpose of ease of expression and that our final results are quite independent of physical considerations.

2. A BASIC PROBLEM

Our study begins with the system

$$\frac{du}{dz}(z)=a(z)u(z)+b(z)v(z), \tag{3.1a}$$

$$-\frac{dv}{dz}(z)=c(z)u(z)+d(z)v(z), \tag{3.1b}$$

defined over some basic interval $\overline{w} \leqslant w \leqslant z \leqslant x \leqslant \overline{x}$. The functions, a, b, c, and d, are required at the moment to be continuous over this interval, although in many important cases they are piecewise continuous. They are scalar functions and need have no relationship to any of the functions denoted by the same or similar letters in the previous chapters. To avoid too many technical questions we shall suppose that for any w and x in the basic interval and for any numbers s_l and s_r the problem defined by (3.1a, b) and the side conditions

$$u(w) = s_l, \tag{3.1c}$$

$$v(x) = s_r, \tag{3.1d}$$

has a unique and well-defined solution. If we wish to revert momentarily to the physical thinking of the past pages we are considering an abstract model of transport in a rod extending from w to x with an input at the left of s_l particles per second and at the right of s_r particles per second. A time independent solution exists and the system is not critical.

Next we define some quite general "reflection" and "transmission" functions:

$$r_r(w,x) = u(x) \qquad \text{when } s_l = 0,\ s_r = 1; \tag{3.2a}$$

$$r_l(w,x) = v(w) \qquad \text{when } s_l = 1,\ s_r = 0; \tag{3.2b}$$

$$t_r(w,x) = v(w) \qquad \text{when } s_l = 0,\ s_r = 1; \tag{3.2c}$$

$$t_l(w,x) = u(x) \qquad \text{when } s_l = 1,\ s_r = 0. \tag{3.2d}$$

Clearly r_r and t_r are precisely the reflection and transmission functions we have been dealing with except that we no longer require that the left end of the system be at zero. The new expressions r_l and t_l are their analogs when the unit input is on the left end of the system instead of the right. Notice that such expressions as $r_r(z_1, z_2)$ are completely meaningful so long as $\overline{w} \leqslant z_1 \leqslant z_2 \leqslant \overline{x}$.

3. THE BASIC FUNCTIONAL EQUATIONS

We choose points z, z_1, z_2 such that $\overline{w} \leqslant w \leqslant z_1 \leqslant z \leqslant z_2 \leqslant x \leqslant \overline{x}$ (Figure 3.1 will prove convenient in the discussion to follow). Then the linearity of the system (3.1) together with the definitions of the r and t functions lead to the following very basic functional equations:

$$u(z) = v(z)\,r_r(z_1, z) + u(z_1)\,t_l(z_1, z), \tag{3.3a}$$

$$v(z) = u(z)\,r_l(z, z_2) + v(z_2)\,t_r(z, z_2). \tag{3.3b}$$

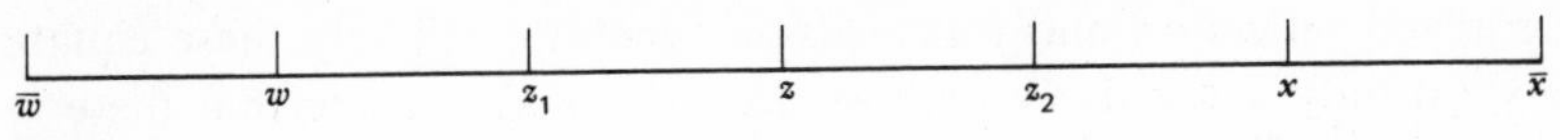

Figure 3.1

We leave the formal derivation to the reader (Problem 1). These results have been obtained in a variety of forms by numerous workers in the field [1–4]. As mentioned, the *validity* of Eqs. (3.3) is assured by the linearity of the system under study. A good *understanding* is perhaps most easily obtained by the type of physical arguments which we have been using. Thus (3.3a) states that the number of right-moving particles passing z each second is due to the input $v(z)$ at the right of the *sub*-rod extending from z_1 to z, resulting in a reflection of particles from that *sub*-rod, plus the contribution caused by the left input of $u(z_1)$ which results in particles being transmitted through that *sub*-rod.

The special case of (3.3) appropriate to the full system under study is

$$u(z)=v(z)r_r(w,z)+s_l t_l(w,z). \tag{3.4a}$$

$$v(z)=u(z)r_l(z,x)+s_r t_r(z,x) \tag{3.4b}$$

We have chosen to write the much more general form (3.3) first; many quite remarkable formulas can be obtained by special choices of z_1 and z_2.

4. SOME APPLICATIONS OF THE RESULTS OF SECTION 3

We observe that (3.3) may be solved for $u(z)$ and $v(z)$:

$$u(z)=\frac{v(z_2)t_r(z,z_2)r_r(z_1,z)+u(z_1)t_l(z_1,z)}{1-r_r(z_1,z)r_l(z,z_2)}, \tag{3.5a}$$

$$v(z)=\frac{u(z_1)t_l(z_1,z)r_l(z,z_2)+v(z_2)t_r(z,z_2)}{1-r_r(z_1,z)r_l(z,z_2)}, \tag{3.5b}$$

or, using the special case (3.4),

$$u(z)=\frac{s_r t_r(z,x)r_r(w,z)+s_l t_l(w,z)}{1-r_r(w,z)r_l(z,x)}, \tag{3.5c}$$

$$v(z)=\frac{s_l t_l(w,z)r_l(z,x)+s_r t_r(z,x)}{1-r_r(w,z)r_l(z,x)}. \tag{3.5d}$$

Thus, the u and v functions may be expressed completely in terms of the generalized reflection and transmission functions. Clearly these equations are valid only if the denominators do not vanish. To see that there is no difficulty, consider, for example, (3.5c). Choose $s_r=0$. If the denominator were zero then (3.5c) could be replaced by $0=s_l t_l(w,z)$. (To see this, carry out the derivation of (3.5c).) But s_l is arbitrary and may surely be chosen nonzero. Thus $t_l(w,z)$ vanishes. This is impossible (see Problem 3).

One wonders if it is possible to somehow eliminate the u and v functions completely using the basic identities (3.3) and hence obtain functional equations involving only r and t. Such results, in simple cases, were first obtained by particle counting type arguments. We shall derive some of them here on a purely analytical basis, and then "verify" one of them by the more classical heuristic reasoning.

First we choose $s_l=0$, $s_r=1$. In (3.5c) replace z by z_1 to obtain

$$u(z_1)=\frac{t_r(z_1,x)r_r(w,z_1)}{1-r_r(w,z_1)r_l(z_1,x)}. \tag{3.6}$$

Next replace z in (3.3a) by x, note that $v(x)=s_r=1$, and replace $u(z_1)$ by its value given in (3.6):

$$u(x)=r_r(z_1,x)+\frac{t_r(z_1,x)r_r(w,z_1)t_l(z_1,x)}{1-r_r(w,z_1)r_l(z_1,x)}. \tag{3.7}$$

Finally, recall that with s_l and s_r as given $u(x)\equiv r_r(w,x)$. Since z_1 is any point in $w\leqslant z\leqslant x$ let us simply call it z. Hence,

$$r_r(w,x)=r_r(z,x)+\frac{t_r(z,x)r_r(w,z)t_l(z,x)}{1-r_r(w,z)r_l(z,x)}. \tag{3.8}$$

As another example we choose s_l and s_r as above and replace z in (3.5d) by z_2:

$$v(z_2)=\frac{t_r(z_2,x)}{1-r_r(w,z_2)r_l(z_2,x)}. \tag{3.9}$$

In (3.3b) we now pick $z=w$,

$$v(w)=u(w)r_l(w,z_2)+\frac{t_r(z_2,x)t_r(w,z_2)}{1-r_r(w,z_2)r_l(z_2,x)}. \tag{3.10}$$

But $u(w)=s_l=0$, $v(w)=t_r(w,x)$ and again z_2 may be any z point between

w and x. Thus

$$t_r(w,x)=\frac{t_r(z,x)t_r(w,z)}{1-r_r(w,z)r_l(z,x)}. \tag{3.11}$$

The number of such identities that can be obtained by making special selections of z's, s values, and the like, seems almost limitless. For some other examples the reader is referred to Problem 4.

These results become especially simple when one is studying the transport model of Section 2, Chapter 1, when the physical parameters $\sigma(z)$, $f(z)$, and $b(z)$ are identically constant. In that case the r and t functions depend upon length alone. Moreover, since the model does not distinguish between left and right we have

$$\begin{aligned} r_l(z_1,z_2)&=r_r(z_1,z_2)=r(z_2-z_1),\\ t_l(z_1,z_2)&=t_r(z_1,z_2)=t(z_2-z_1). \end{aligned} \tag{3.12}$$

Hence, choosing $w=0$ to obtain further agreement with the notation of Chapter 1, we obtain from (3.8) and (3.11)

$$r(x)=r(x-z)+\frac{t^2(x-z)r(z)}{1-r(z)r(x-z)}, \tag{3.13a}$$

$$t(x)=\frac{t(x-z)t(z)}{1-r(z)r(x-z)}. \tag{3.13b}$$

The choice $z=\frac{1}{2}x$ is particularly interesting:

$$r(x)=r\left(\frac{x}{2}\right)+\frac{t^2(x/2)r(x/2)}{1-r^2(x/2)}, \tag{3.14a}$$

$$t(x)=\frac{t^2(x/2)}{1-r^2(x/2)}. \tag{3.14b}$$

We notice that the criticality condition now becomes $r(\frac{1}{2}x)=1$. That is, a rod with constant physical parameters is critical when a rod of half its length reflects exactly the same number of particles as are injected into it. Of course, none of these results is of such a nature that it cannot be obtained without the use of the functional equations we have derived, for the rod case with constant parameters can, after all, be solved explicitly. The solution will, in general, be in terms of trigonometric functions. Equations (3.13) reduce in this instance to trigonometric identities—often very messy ones. Much more interesting results are obtained when the functional equation approach is used on matrix systems (see Problem 6) or in the case of equations whose coefficients are periodic functions of z (see Chapter 10).

Equations (3.14) have still another use. Notice that if r and t are known at $\frac{1}{2}x$ they can be very easily computed at x. Continuing this argument, we can readily find their values at $2x, 4x, \ldots, 2^n x$. The use of this simple observation can considerably reduce the amount of integration of differential equations required in many problems. Its value in realistic transport problems was first noted by Van de Hulst [5] who refers to it as the *method of doubling*. The idea can also be extended to the periodic problems alluded to in the previous paragraph.

Before leaving this topic, let us obtain Eq. (3.13a) by particle counting. Recall that z may be any point internal to the rod and that the left end has been chosen as zero. (See Figure 3.2.)

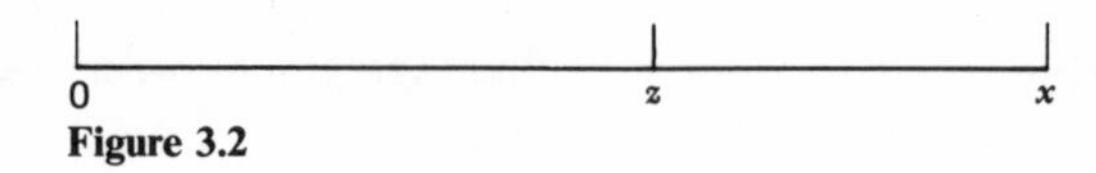

Figure 3.2

Consider a particle entering at the right end. It "sees" ahead of it *two sub*-rods, one extending from z to x, the other from zero to z. Some of the progeny of this injected particle will arise from interactions only in the *sub*-rod between z and x. Such interactions will produce $r(x-z)$ particles per second at the right end of the rod, and these particles will contribute to $r(x)$. However, other particles will be transmitted through the interval (z,x) and form a source for the *sub*-rod which extends from zero to z. The strength of this source is $t(x-z)$. This results in particles being reflected by this *sub*-rod, the total being $t(x-z)r(z)$ per second. This is now a source for the interval (z,x), and results in a total of $t(x-z)r(z)t(x-z)$ particles each second emergent at x. It also results in a total of $t(x-z)r(z)$ $r(x-z)$ being reflected each second back into the *sub*-rod extending from zero to x. It in turn reflects particles back, and so on. It is clear that we must now keep complete and careful account of all multiple reflections. When we do so we find an infinite series expression for $r(x)$:

$$\begin{aligned} r(x) = r(x-z) &+ t(x-z)r(z)t(x-z) \\ &+ t(x-z)r(z)r(x-z)r(z)t(x-z) \\ &+ t(x-z)r(z)r(x-z)r(z)r(x-z)r(z)t(x-z) \\ &+ \cdots . \end{aligned} \tag{3.15}$$

If we now assume that the infinite series converges, and it will if and only if $|r(z)r(x-z)|<1$, then we readily obtain the previous equation (3.13a).

It is obvious that the functional equation approach based upon Eqs. (3.3) is vastly superior to this particle counting method. Such heuristic reasoning can easily lead to errors, especially in more complicated cases. Moreover, the requirement that $|r(z)r(x-z)|<1$ seems here to be quite an artificial one. Nevertheless, it is valuable to know the true physical significance of equations like (3.13a), and the reader is urged to obtain some experience in this kind of reasoning (see Problems 7–9).

5. DIFFERENTIAL EQUATIONS VIA FUNCTIONAL EQUATIONS

We shall now show how to use the functional equations obtained in the last two sections to derive differential equations satisfied by the r and t functions. Our first method is presented more or less as a matter of historical interest. It was actually originally used on Eq. (3.13a) to obtain a differential equation for $r(x)$. We shall use it on Eq. (3.8). Then we shall show that the same result may be obtained much more elegantly from the very basic Eq. (3.3).

In (3.8) choose $z=w+\Delta$ to get

$$\frac{r_r(w,x)-r_r(w+\Delta,x)}{\Delta}=\frac{t_r(w+\Delta,x)t_l(w+\Delta,x)}{1-r_r(w,w+\Delta)r_l(w+\Delta,x)}\cdot\frac{r_r(w,w+\Delta)}{\Delta}. \tag{3.16}$$

Obviously we are eventually going to allow Δ to approach zero. Since the various r and t functions are continuous in all of their arguments, the most troublesome term will evidently be $r_r(w,w+\Delta)/\Delta$. To study its behavior we examine a system which extends from w to $w+\Delta$, has zero input on the left and unit input on the right. (See Figure 3.3.) If we write out Eq. (3.1a) in finite difference form we obtain (again making full use of continuity)

$$u(w+\Delta)-u(w)=a(w)u(w)\Delta+b(w)v(w+\Delta)\Delta+o(\Delta). \tag{3.17}$$

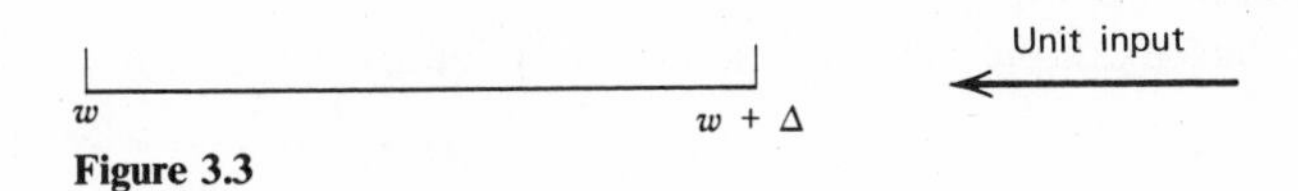

Figure 3.3

But $u(w)=0$, $v(w+\Delta)=1$, and $u(w+\Delta)=r_r(w,w+\Delta)$, by definition. Thus

$$\frac{r_r(w,w+\Delta)}{\Delta}\to b(w) \qquad \text{as } \Delta\to 0. \tag{3.18}$$

Taking the limit in (3.16) then yields at once

$$r_{r,1}(w,x) = -t_r(w,x)t_l(w,x)b(w), \tag{3.19}$$

where we have used the numerical subscript notation to denote partial differentiation. Observe that Eq. (3.19) is of a different nature from other differential equations that we have previously encountered for r and t functions. The formula tells us how r_r varies when the *left* end of the system is perturbed. In the previous chapters the left end of the rod has been fixed in all cases dealt with in detail.

A much neater way of deriving (3.19) involves starting directly with (3.3a). We continue to investigate a system which extends from w to x, with zero input at w and unit input at x. In Eq. (3.3a) then, u and v really depend upon w and x, and these inputs, although we have not indicated that in our notation. On the other hand, z_1 and z_2 are in no way dependent, either explicitly or implicitly, upon any of these quantities. They are only required to satisfy the conditions $w \leqslant z_1 \leqslant z \leqslant z_2 \leqslant x$, where, of course, $\overline{w} \leqslant w$ and $x \leqslant \bar{x}$. Bearing these dependencies in mind, we differentiate (3.3a) with respect to z_1:

$$0 = v(z)r_{r,1}(z_1,z) + u_1(z_1)t_l(z_1,z) + u(z_1)t_{l,1}(z_1,z). \tag{3.20}$$

But $u_1(z_1)$ can be obtained directly from (3.1a):

$$u_1(z_1) = a(z_1)u(z_1) + b(z_1)v(z_1). \tag{3.21}$$

Now we select $z_1 = w$, $z = x$, and recall that

$$u(w) = 0, \qquad v(w) = t_r(w,x), \qquad v(x) = 1. \tag{3.22}$$

From Eqs. (3.20)–(3.22) we readily calculate:

$$r_{r,1}(w,x) = -t_r(w,x)t_l(w,x)b(w), \tag{3.23}$$

in agreement with Eq. (3.19).

As another example of the use of (3.3) let us set $z_1 = w$, $u(w) = 0$ in (3.3a):

$$u(z) = v(z)r_r(w,z). \tag{3.24}$$

If we compare (3.24) with Eq. (2.40), identify $\rho(z)$ with r_r and think of w, the left end of the system as being at zero, we will note that the two equations differ only by the function $w(z)$. The role of that function in the development of Section 8 of Chapter 2 was solely to care for the inhomogeneous terms $s^+(z)$ and $s^-(z)$ arising in Eqs. (2.38). Our current system (3.1) contains no inhomogeneous terms. It is readily verified that the manipulations of Section 8, Chapter 2, when (3.24) is used in (3.1) will yield the usual r equation

$$r_{r,2}(w,z) = b(z) + \{a(z) + d(z)\} r_r(w,z) + c(z) r_r^2(w,z). \tag{3.25}$$

A very extensive set of differential equations relating the r and t functions can be obtained by devices such as these. We leave the derivation of some others to the reader (see Problem 10; see also [6]). Such collections of differential equations are convenient for the analysis of certain questions of practical interest. (See, for example, Chapter 10.) Initial conditions may always be obtained directly from the definitions of the various r and t functions, Eqs. (3.2).

6. SUMMARY

In this chapter we have discussed perhaps the most basic equations of invariant imbedding, Eqs. (3.3). From these we have indicated how extensive classes of functional equations and differential equations relating various r and t functions can be derived. Although we have concentrated on the homogeneous system (3.1), the insights obtained now make it clear how such artificial appearing transformations as that used in Section 8 of Chapter 2 can be found when nonhomogeneous problems arise.

For convenience in exposition we have studied only the scalar (one-state) system (3.1). For vector systems, much of the basic reasoning remains unchanged, but one must naturally exercise great care in handling matters of commutativity, inverses (in place of reciprocals), and the like. Again we leave many of these matters as problems. (See Problem 5.)

We have now presented the fundamentals of the invariant imbedding technique. In the chapters that follow we shall begin to make use of these ideas in special situations of interest to engineers, physicists, applied mathematicians, and numerical analysts. It is our hope that anyone who has mastered the basics thus far presented can focus on those chapters of particular interest to him if he so wishes.

PROBLEMS

1. Give a rigorous derivation of Eqs. (3.3). Also give the matrix analogs of both (3.3) and (3.4). Use both physical and rigorous analytical reasoning.

2. Discuss the possibility of replacing the requirement of continuity on a, b, c, d (Eq. 3.1) with a lesser restriction. Determine if this new assumption is satisfactory for the work of the entire chapter.

3. Prove that the t functions are never zero. What is the analog in the matrix case?

4. Obtain analogs of (3.8) for r_l and of (3.11) for t_l.

5. Generalize (3.5) to matrix systems. Give an argument that shows that the inverse matrix which replaces the denominator of (3.5) is nonsingular. Obtain analogs of both (3.8) and (3.11) in the matrix case.

6. Generalize Eqs. (3.14) to the matrix case and study criticality in n-state physical systems in this way.

7. Generalize the particle counting approach used in obtaining (3.15) to the case of nonconstant $f(z)$ and $b(z)$ and to scalar equations of the form (3.1).

8. Generalize the particle-counting approach of Problem 7 to the case of n-state systems.

9. Use the particle counting ideas of the previous two problems to obtain functional equations for the transmission functions.

10. A full set of differential equations for the r and t functions is (we write d/dz instead of $\partial/\partial z$ since the arguments y and x are treated as constants):

$$\frac{d}{dz}r_r(y,z)=b(z)+[a(z)+d(z)]r_r(y,z)+c(z)r_r^2(y,z),$$

$$r_r(y,y)=0;$$

$$\frac{d}{dz}t_r(y,z)=[d(z)+c(z)r_r(y,z)]t_r(y,z),$$

$$t_r(y,y)=1;$$

$$\frac{d}{dz}r_l(y,z)=c(z)t_l(y,z)t_r(y,z),$$

$$r_l(y,y)=0;$$

$$\frac{d}{dz}t_l(y,z)=t_l(y,z)[a(z)+c(z)r_r(y,z)],$$

$$t_l(y,y)=1;$$

$$-\frac{d}{dz}r_r(z,x)=b(z)t_r(z,x)t_l(z,x),$$

$$r_r(x,x)=0;$$

$$-\frac{d}{dz}t_r(z,x)=[d(z)+b(z)r_l(z,x)]t_r(z,x),$$

$$t_r(x,x)=1;$$

$$-\frac{d}{dz}r_l(z,x)=c(z)+[a(z)+d(z)]r_l(z,x)+b(z)r_l^2(z,x),$$

$$r_l(x,x)=0;$$

$$-\frac{d}{dz}t_l(z,x)=[a(z)+b(z)r_l(z,x)]t_l(z,x),$$

$$t_l(x,x)=1.$$

(See, for example, R. C. Allen, Jr., and G. M. Wing, "A Numerical Algorithm Suggested by Problems of Transport in Periodic Media," *J. Math. Anal. Appl.* **29**, 1970, 141–157.)

Obtain these equations from the functional equation approach. Also obtain their matrix analogs.

11. Consider the scalar problem

$$u'(z)=a(z)u(z)+b(z)v(z)+s^{+}(z),$$

$$-v'(z)=c(z)u(z)+d(z)v(z)+s^{-}(z).$$

Try to write functional equations analogous to (3.3) using one or more additional terms to account for the inhomogenity. Consider the possibility of getting the results of Chapter 2, Section 8 in this way—a somewhat less ad hoc method than used in that chapter. Generalize to the matrix case.

12. Since the background from which the method of invariant imbedding has emerged is basically transport theory, the problems studied have ordinarily had conditions on the u-like functions imposed at one end of the interval and conditions on the v-like functions imposed at the other end. This is by no means necessary. Use the fundamental equation (3.3) to show how to solve a scalar (u,v) problem when the value of u is assigned at both ends of the interval. (See Eq. (7.63) of Chapter 7.) Generalize this to the matrix case. Consider both scalar and matrix problems when v is specified at both ends.

13. Pursue the ideas of Problem 12 further and attempt to formulate the most general boundary conditions which can be handled by the imbedding methods through the use of (3.3). (See also Problem 8, Chapter 2.)

14. As a physical example of the above consider a transport problem in which a reflecting medium is adjacent to the medium under study. Thus the value of u at that end point is given in terms of the value of v at that point. More specifically, suppose in a scalar problem the condition at $z=0$ is $u(0)=\beta v(0)$. That is, the reflecting medium on the left returns on the average β particles to the system for each particle emerging at the left. Let the condition at the right end remain $v(x)=1$. Show how to resolve this problem by invariant imbedding. Consider the matrix analog, keeping the physics in mind.

15. Consider the scalar differential equation $dy/dt=g(y)$, $y(0)=c$. Suppose this possesses a unique solution for all $t\geqslant 0$ and all values of c. Write $y(t)=y(t,c)$. Show that this function satisfies the functional equation

$$y(t+s,c)=y(t,y(s,c)).$$

Do this from two viewpoints:

a. Both sides of the functional equation represent a procedure for determining the "state" of a system at time $t+s$. Thus the equation is really an analytical representation of the "law of causality."

b. The two sides of the functional equation are equal as a result of the assumption that the original differential equation has a *unique* solution.

16. By appropriate choice of g in Problem 15 obtain at once the result $e^{t+s}=e^{t}e^{s}$.

17.* Using reasoning similar to that employed in the above two problems derive the addition formulas for the sine and cosine functions by studying the differential equation

$$y''+y=0, \qquad y(0)=c_1, \qquad y'(0)=c_2.$$

18. Show that if $X(t)$ is a differentiable matrix function satisfying the functional equation $X(t+s)=X(t)X(s)$ for all real t and s with $X(0)=I$, the identity matrix, then $X(t)=e^{At}$ for some constant matrix A. Do this by differentiating both sides of the functional equation with respect to both t and s and comparing results.

19.* Show that the result of Problem 18 holds even when X is merely required to be continuous. (This is a classical result of Polya.)

20. Return to Problem 15 and try to generalize the result to matrix differential equations. Thus derive the result $e^{A(t+s)}=e^{At}e^{As}$. Deduce that e^{At} is never singular and that $(e^{At})^{-1}=e^{-At}$.

21.* As another example of an interesting functional equation consider the matrix transformation

$$f(A,T)=T^{-1}\left(AT-\frac{dT}{dt}\right),$$

where A and T are appropriately behaved matrix functions of t. By considering the differential equation $dx/dt=A(t)x$, $x(0)=c$, and the transformation $x=Ty$, prove that

$$f(A,ST)=f(f(A,T),S).$$

22.* Consider the multipoint boundary value problem

$$L(y)=y'''+p_1(t)y''+p_2(t)y'+p_3(t)y=0,$$

$$y(0)=0, \qquad y(a)=0, \qquad y(b)=1.$$

Suppose that this problem is soluble for all a and b, $0<a<b\leqslant b^*$. Write $v(b)=y'(b)$ and $w(b)=y''(b)$. Show that

$$v(b)=\frac{\begin{vmatrix} u_2'(b) & u_3'(b) \\ u_2(a) & u_3(a) \end{vmatrix}}{\begin{vmatrix} u_2(b) & u_3(b) \\ u_2(a) & u_3(a) \end{vmatrix}},$$

$$w(b)=\frac{\begin{vmatrix} u_2''(b) & u_3''(b) \\ u_2(a) & u_3(a) \end{vmatrix}}{\begin{vmatrix} u_2(b) & u_3(b) \\ u_2(a) & u_3(a) \end{vmatrix}},$$

where u_2 and u_3 are the solutions of $L(u)=0$ determined by $u_2(0)=0$, $u_2'(0)=1$, $u_2''(0)=0$, and $u_3(0)=0$, $u_3'(0)=0$, and $u_3''(0)=1$.

23.* In Problem 22 set

$$z(t)=\begin{vmatrix} u_2(t) & u_3(t) \\ u_2(a) & u_3(a) \end{vmatrix},$$

so that $v(b)=z'(b)/z(b)$ and $w(b)=z''(b)/z(b)$. Thus at $t=b$, $z'(b)-v(b)z(b)=0$. Differentiate with respect to b and obtain

$$z''-v'z-vz'=0,$$

$$z'''-v''z-2v'z'-vz''=0,$$

where the prime henceforth denotes differentiation with respect to b. Show that this leads to

$$\begin{vmatrix} 0 & 1 & -v \\ 1 & -v & -v' \\ (p_1+v) & (p_2+2v') & (p_3+v'') \end{vmatrix}=0,$$

where all evaluations are at b. Derive a similar result for w. Obtain the initial condition at $b=a$.

Now consider the relationship between these ideas and those of invariant imbedding. Notice that the right-hand end point b has become the variable. Can you analyze the problem more directly using the imbedding concepts?

24.* Given the diffusion equation $u_t=u_{xx}$. Set $w=u_x/u$. (Compare this transformation with standard invariant imbedding transformations.) Derive Burger's equation

$$\begin{vmatrix} -w & 1 & 0 & 0 \\ -w_t & 0 & -w & 1 \\ -w_x & -w & 1 & 0 \\ -w_{xx} & -2w_x & -w & 1 \end{vmatrix}=0.$$

25.* Use the device of Problem 24 to obtain an equation for $v=u_x/u$ when $u_t=u_{xxx}$.

26.* Let $y(t)$ satisfy $y''+p(t)y'+q(t)y=0$. Suppose $\phi_1(t)$ and $\phi_2(t)$ are two given functions. Consider the solubility of the problem

$$\int_0^x y(t)\phi_1(t)\,dt=b_1, \qquad \int_0^x y(t)\phi_2(t)\,dt=b_2,$$

where b_1 and b_2 are given constants. To do this let y_1 and y_2 be linearly independent solutions of the differential equation. Show that the existence and uniqueness of the solution to the foregoing problem is dependent upon the nonvanishing of the determinant

$$\Delta(y,\phi)=\begin{vmatrix} \int_0^x y_1(t)\phi_1(t)\,dt & \int_0^x y_1(t)\phi_2(t)\,dt \\ \int_0^x y_2(t)\phi_1(t)\,dt & \int_0^x y_2(t)\phi_2(t)\,dt \end{vmatrix}.$$

27.* Show that the determinant in Problem 25 can be written in the form

$$\Delta(y,\phi)=\frac{1}{2}\int_0^x\int_0^x \begin{vmatrix} y_1(t_1) & y_1(t_2) \\ y_2(t_1) & y_2(t_2) \end{vmatrix} \cdot \begin{vmatrix} \phi_1(t_1) & \phi_1(t_2) \\ \phi_2(t_1) & \phi_2(t_2) \end{vmatrix} dt_1\,dt_2.$$

Now show that the domain of integration may be taken as $0 \leqslant t_1 \leqslant t_2 \leqslant x$.

28.* Using the results of the above two problems discuss the solubility of $y''+y=0$ subject to

$$\int_0^x y(t)e^{\lambda_1 t}\,dt=b_1, \qquad \int_0^x y(t)e^{\lambda_2 t}\,dt=b_2.$$

(For more general results see R. Bellman, A Note on the Identification of Linear Systems, *Proc. Amer. Math. Soc.*, **17**, 1966, 68–71.)

29.* The class of questions posed in the above problems may also be approached from the viewpoint of invariant imbedding. First, convert the differential equation of Problem 26 into a system of two first order differential equations. Now employ Eqs. (3.5). Obviously there is no information concerning the boundary conditions. Show how to obtain this information by using the integral conditions of Problem 26. Specialize to the case of Problem 28, using the decomposition

$$\frac{du}{dt}=v, \qquad -\frac{dv}{dt}=u,$$

where $u=y$.

In what ways is the invariant imbedding approach inferior to the classical one described in the problems? Does it have any advantages?

30. Consider the integral equation

$$\phi(z)=1+\int_0^x \exp(k|z-t|)\phi(t)\,dt.$$

By differentiation with respect to z show that $\phi(z)$ satisfies a simple second order differential equation with constant coefficients. Determine the boundary conditions.

31. Convert the differential equation of Problem 30 into a system of (u,v)-like equations with a boundary condition on u at zero and a boundary condition on v at x. Now employ the method of invariant imbedding, thus obtaining a set of r and t equations. Note that you have reduced the solution of an integral equation to an invariant imbedding problem. (For extensive results, see Chapter 12 of this book.)

32.* Given the problem $y''+a(t)y=0$, $y(0)=c_1$, $y'(x)=c_2$. Observe that this can be reduced to a system of the sort we have been studying by invariant imbedding. Consider the following iterative scheme. Let $y_0(0)=c_1$ and $y_0'(0)=b_0$, b_0 arbitrary. Calculate $y_0(x)=b_1$. Now set $y_1(x)=b_1$ and $y_1'(x)=c_2$. Compute $y_1'(0)=b_2$. Set $y_2(0)=c_1$ and $y_2'(0)=b_2$, and so on. Does this procedure converge? (See R. Bellman and T. Brown, "On the Iterative Solution of Two-Point Boundary Value Problems," *Boll. U.M.I.* **16**, 1961, 145–149; and R. Bellman and T. Brown, "On the Computational Solution of Two-Point Boundary Value Problems," *Boll. U.M.I.* **19**, 1964, 121–123.)

REFERENCES

1. R. C. Allen, Jr., "Functional Relationships for Fredholm Integral Equations Arising from Pseudo-Transport Problems," *J. Math, Anal. Appl.* **30**, 1970, 48–78.
2. R. M. Redheffer, "On the Relation of Transmission-line Theory to Scattering and Transfer," *J. Math. Phys.* **41**, 1962, 1–41.
3. R. W. Preisendorfer, *Radiative Transfer on Discrete Spaces*, Pergamon, Oxford, 1965.
4. E. D. Denman, "Invariant Imbedding and Linear Systems," in *Invariant Imbedding* (R. Bellman and E. D. Denman, Eds.), Springer Verlag, Berlin, 1971.
5. H. C. Van de Hulst, *A New Look at Multiple Scattering*, NASA Institute for Space Studies, Goddard Space Flight Center, January 1963.
6. W. T. Reid, "Solutions of a Riccati Matrix Differential Equation as Functions of Initial Values," *J. Math. Mech.* **8**, 1959, 221–230.

4

EXISTENCE, UNIQUENESS, AND CONSERVATION RELATIONS

1. INTRODUCTION

The results of these early chapters suggest that the method of invariant imbedding, coupled with physical insight, may provide a way of obtaining information concerning the existence and uniqueness of solutions to certain classes of linear two-point boundary value problems. In this chapter we pursue this matter and verify that these hopes are indeed justified. In the text we consider the case of constant coefficients, while that of variable coefficients is left to the exercises at the end of the chapter.

Let us turn first to Eqs. (1.32) and the physics surrounding them. One certainly conjectures that if the collision process is such that no more particles emerge (on the average) from a collision than enter into that collision, then that physical system cannot become critical. Equivalently, the corresponding two-point value problem will have a solution no matter what the size of the system. Such an idea should generalize to equations which do not arise directly from transport processes, but which are in some sense models of generalized transport phenomena.

It turns out to be convenient and instructive to break our investigation into two parts. First, we shall consider analogs of conservative systems, those in which a total of exactly one particle (on the average) emerges from each collision. Having verified our conjecture in that case we shall turn to the nonconservative situation in which no more than one particle (again on the average) is produced in the collision interaction. The conservative case is obviously a special instance of this. However, successful analysis of the nonconservative problem requires the introduction of an important new function, a dissipation function [1,2] which has not been treated heretofore. It is identically zero in the conservative case.

We make no effort here to investigate problems in which the number of particles is actually increased by the interaction process. Intuitively it is clear that such systems can become critical and thus the size in general determines the existence or nonexistence of the solution to the corresponding two-point boundary value problem. In special cases, however, it is possible for such a configuration to remain subcritical despite its length (see Problem 1). To the best of our knowledge this type of phenomenon has not been considered in detail.

Finally, it should be remarked again that the analysis given here is only for systems in which all parameters are independent of position. Thus we study only systems of equations with constant coefficients. The ideas, however, may rather easily be applied to the more general case where the coefficients are variable (see Problems 2, 10–12).

2. THE "PHYSICS" OF THE CONSERVATIVE CASE AND ITS GENERALIZATIONS

From the definitions of f_{ij} and b_{ij} (Chapter 1, Section 7), we see that the condition for conservative collision (sometimes called pure scattering) is

$$\sum_{i=1}^{n} (f_{ij}+b_{ij})=1, \qquad j=1,2,\ldots,n; \tag{4.1}$$

or, equivalently,

$$\sigma_j \sum_{i=1}^{n} [(f_{ij}-\delta_{ij})+b_{ij}]=0, \qquad j=1,2,\ldots,n. \tag{4.2}$$

Let us translate this into terms appropriate to the generalized Eqs. (2.47) which we now write in the source-free form

$$\begin{aligned} \frac{du}{dz} &= Au+Bv, \\ -\frac{dv}{dz} &= Cu+Dv. \end{aligned} \tag{4.3}$$

Here the coefficients A, B, and so on, are henceforth to be considered as *constant* matrices. No confusion should result from the apparent double use of the symbol B since almost all of our mathematical arguments deal with Eq. (4.3). The analog of (4.2) is

$$\sum_{i=1}^{n}(a_{ij}+c_{ij})=0, \qquad \sum_{i=1}^{n}(b_{ij}+d_{ij})=0, \qquad j=1,2,\ldots,n. \tag{4.4}$$

We shall also require

$$a_{ij}\geqslant 0 \text{ and } d_{ij}\geqslant 0 \text{ for } i\neq j;\ b_{ij}\geqslant 0 \text{ and } c_{ij}\geqslant 0 \text{ for all } i,j. \tag{4.5}$$

We assert that under these assumptions the function $R_r(a,x)$ exists for all a and all x, $a\leqslant x$. In view of the fact that the corresponding functions R_l, T_r, and T_l are obtainable from R_r by the solution of a set of linear differential equations (see Problem 10, Chapter 3), it follows that all of these functions exist and are unique regardless of the values of a and x. The functional equations (3.4) when generalized to the matrix case (see Problems 5, Chapter 3) then assure the existence and uniqueness of the solution to any problem (4.3) subject to any boundary conditions

$$u(a)=u_a, \qquad v(b)=v_b, \qquad a<b. \tag{4.6}$$

To verify all of these assertions it is convenient to examine still another derivation of the R_r function. This derivation will bring into focus several results from classical differential equation theory which will be useful in achieving our ultimate goal.

3. ANOTHER DERIVATION OF THE REFLECTION FUNCTION

We begin by replacing the system (4.3) by a matrix system

$$\begin{aligned}\frac{dU}{dz}&=AU+BV,\\ -\frac{dV}{dz}&=CU+DV.\end{aligned} \tag{4.7}$$

Here $U(z)$ and $V(z)$ are to be regarded as square matrices. Let $U_1(z)$ and $V_1(z)$ solve (4.7) subject to the *initial* conditions

$$U_1(a)=0, \qquad V_1(a)=I. \tag{4.8}$$

That such functions exist uniquely is a consequence of the classical existence theorems for such problems. Now if $R_r(a,x)$ exists, it is easy to see that (note Eq. (3.24))

$$U_1(x) = R_r(a,x)V_1(x) \tag{4.9}$$

or

$$R_r(a,x) = U_1(x)V_1^{-1}(x). \tag{4.10}$$

But the condition $V_1(a) = I$ and the fact that V_1 is a continuous function of z implies that for z sufficiently near a the inverse matrix $V_1^{-1}(z)$ does indeed exist. Thus Eq. (4.10) serves to *define* $R_r(a,x)$ for all x sufficiently close to a and to the right of a, say for $a \leqslant x \leqslant x_0$. Finally, all functions involved are differentiable on this interval.

Using these observations we may now obtain a differential equation for R_r*:

$$\begin{aligned}
\frac{d}{dx}R_r(a,x) &= \left[\frac{d}{dx}U_1(x)\right]V_1^{-1}(x) + U_1(x)\frac{d}{dx}V_1^{-1}(x) \\
&= \left[\frac{d}{dx}U_1(x)\right]V_1^{-1}(x) + U_1(x)\left[-V_1^{-1}(x)\frac{dV_1(x)}{dx}V_1^{-1}(x)\right] \\
&= [AU_1(x) + BV_1(x)]V_1^{-1}(x) \\
&\quad + U_1(x)V_1^{-1}(x)[CU_1(x) + DV_1(x)]V_1^{-1}(x) \\
&= B + AR_r(a,x) + R_r(a,x)D + R_r(a,x)CR_r(a,x).
\end{aligned} \tag{4.11}$$

This result is in agreement with those found in earlier chapters. Our current approach makes it clear that all reasoning is valid in the interval $a \leqslant x \leqslant x_0$. There is no assurance, however, that (4.11) is valid for *all* x. It is our purpose now to show that this is indeed the case. To accomplish this we make some further use of physical analogies, and then put our thinking on a rigorous basis.

4. SOME CONSERVATION RELATIONS

It is convenient to return to Eq. (4.4). In the physical context the assumption of pure scattering implied by this equation also means that no particles may be lost within the system; a total of one particle entering the system must result in a total of one particle leaving the system. Mathemati-

*In discussions in which one argument of an R- or T-like function remains fixed we shall often write the derivative with respect to the other argument as an ordinary rather than a partial derivative.

cally this is most easily expressed by introducing the "column-summing matrix" M:

$$M = \begin{pmatrix} 1 & 1 & 1 & \cdots & 1 \\ 0 & 0 & 0 & \cdots & 0 \\ \vdots & & & & \\ 0 & 0 & 0 & \cdots & 0 \end{pmatrix}. \tag{4.12}$$

The relationship (4.4) may then be written

$$M(A+C)=0, \qquad M(B+D)=0, \tag{4.13}$$

while the statement that no particles may be lost within the system becomes

$$M(R_r+T_r)=M. \tag{4.14}$$

We shall next prove that (4.14) actually holds in $a \leqslant x \leqslant x_0$.

We recall that T_r satisfies the relation

$$\frac{d}{dx}T_r(a,x)=T_r(a,x)[D+CR_r(a,x)], \tag{4.15}$$

a fact that may either be accepted from previous work, or derived using arguments similar to those of Section 3 (see Problem 3). The latter derivation makes it clear that there is a nonempty interval, which we may as well again take to be $a \leqslant x \leqslant x_0$, on which this equation holds. Of course,

$$R_r(a,a)=0, \qquad T_r(a,a)=I, \tag{4.16}$$

so that (4.14) is indeed correct at $x=a$. We must now verify it for $a < x \leqslant x_0$.

Using the fact that M is a constant matrix, we calculate

$$\begin{aligned} \frac{d}{dx}M(R_r+T_r) &= M(B+AR_r+R_rD+R_rCR_r)+M(T_rD+T_rCR_r) \\ &= M(R_r+T_r-I)D+M(R_r+T_r-I)CR_r \\ &\quad + M(A+C)R_r+M(B+D) \\ &= M(R_r+T_r-I)D+M(R_r+T_r-I)CR_r. \end{aligned} \tag{4.17}$$

Setting

$$Z = M(R_r + T_r - I), \tag{4.18}$$

we find then that Z satisfies

$$\frac{dZ}{dx} = Z(D + CR_r), \qquad Z = 0 \text{ at } x = a. \tag{4.19}$$

This equation, valid on $a \leqslant x \leqslant x_0$, has the unique solution $Z = 0$. Thus (4.14) is verified in this interval.

The problem that still remains is to overcome the restriction that x lie *only* in this interval. To resolve this question we shall turn to a common device of differential equation theory, making strong use of the relationship

$$M(R_r(a,x) + T_r(a,x) - I) = 0, \qquad a \leqslant x \leqslant x_0. \tag{4.20}$$

5. PROOF OF EXISTENCE IN THE CONSERVATIVE CASE

We shall show first that in $a \leqslant x \leqslant x_0$ the solution to (4.11) is nonnegative. That is, each element of the R_r matrix is nonnegative. To see this, we first write an integral equation equivalent to (4.11). Using the fact that A, B, and so on, are constant matrices we find

$$R_r(a,x) = \int_a^x e^{+A(x-t)}[B + R_r(a,t)CR_r(a,t)]e^{+D(x-t)}\,dt. \tag{4.21}$$

A standard method of analyzing this problem is to introduce the sequence $\{R_{r,n}\}$ defined recurrently by

$$R_{r,n+1}(a,x) = \int_a^x e^{+A(x-t)}[B + R_{r,n}(a,t)CR_{r,n}(a,t)] \cdot e^{+D(x-t)}\,dt, \qquad n = 0, 1, 2, \ldots, \tag{4.22}$$

$$R_{r,0}(a,x) = 0.$$

It is well known that if x_0 is small enough this sequence converges to the unique solution of (4.11); see Problem 4.

We next need a lemma.

LEMMA. If H is a matrix with nonnegative off-diagonal elements, $h_{ij} \geqslant 0$, $i \neq j$, then e^H is a nonnegative matrix.

PROOF. Choose the number k so that $k+h_{ii}\geqslant 0$ for all i. Then write

$$e^{H}=e^{(H+kI)}e^{-kI}. \tag{4.23}$$

Since both matrices on the right are nonnegative the result is immediate.

Applying the lemma to (4.22) and recalling that B and C are nonnegative while A and D have nonnegative off-diagonal elements we see that the approximating sequence $R_{r,n}$ has all nonnegative elements. Obviously the same is true of the limit.

If we apply the same kind of reasoning to Eq. (4.15) we find that the matrix T_r is also nonnegative. Equation (4.20) now reveals that all elements of the R_r and T_r matrices are in addition no greater than unity.

All the essential ingredients for the existence proof have now been obtained, and we merely sketch the remainder, referring the reader to standard treatises on differential equations for the details, ([3]). Suppose the solution to the R_r equation fails to exist at the point x_1 but does exist for all x values such that $a\leqslant x<x_1$. The boundedness of the individual elements of the R_r matrix allows one to prove that $\lim_{x\to x_1^-} R_r(a,x)$ exists. Furthermore, this limiting value may be taken as the value of $R_r(a,x_1)$. But now the above analysis may be repeated using $R_r(a,x_1)$ as the initial value at the new initial point x_1. Hence the solution of the R_r equation is extended to the right of the point x_1 at which it supposedly failed to exist. This is a contradiction, and therefore there can be no such point as x_1. Thus the solution exists everywhere and we have completed the proof.

6. THE NONCONSERVATIVE CASE: THE DISSIPATION FUNCTION

We turn to the problem alluded to in Section 1, that in which physically the number of particles out of a collision interaction averages less than one for each moving particle involved in the collision. A process of "absorption" results in the loss of particles, and it is therefore reasonable to suppose that the total number of particles emergent from the system is less than the number of particles entering the system. When this "absorption" process is abstracted to the case of Eq. (4.3) it is to be expected that the corresponding two-point boundary value problem will always have a solution—there can be no phenomenon analogous to criticality. This we shall now prove.

It is helpful to return to the one dimensional transport case and once again engage in particle counting. By so doing we shall arrive at an equation for a new function, the "dissipation" function [1,2], which determines the number of particles actually lost in the system. This function is easily generalized to the case of systems of equations, again by particle counting. It is through the analysis of the dissipation function that we shall obtain our result. We shall also observe that the physical meaning of the dissipation function is not important in the existence proof. Only the differential equation that it satisfies really plays a role. Thus, while the particle counting technique is useful in leading us to an important function and an equally important differential equation, the counting procedure in itself is not necessary. Hence the reasoning is on a rigorous analytic basis and is completely independent of physical considerations.

Consider now the one dimensional transport case of Chapter 1 and write

$$\begin{aligned} \frac{du}{dz} &= \sigma\{(f-1)u(z)+bv(z)\}, \\ -\frac{dv}{dz} &= \sigma\{bu(z)+(f-1)v(z)\}, \end{aligned} \tag{4.24}$$

subject to

$$u(a)=0, \qquad v(x)=1. \tag{4.25}$$

Define the dissipation function

$l_r(a,x)=$ the average number of particles lost *within* the system each second due to the input described by Eq. (4.25). (4.26)

Consider the system augmented to extend to $z=x+\Delta$. Particle counting then yields

$$\begin{aligned} l_r(a,x+\Delta) &= \sigma\Delta(1-f-b)+\sigma\Delta f l_r(a,x) \\ &\quad +(1-\sigma\Delta)l_r(a,x)+(1-\sigma\Delta)r_r(a,x)\sigma\Delta(1-f-b) \\ &\quad +(1-\sigma\Delta)r_r(a,x)\sigma\Delta b l_r(a,x)+o(\Delta). \end{aligned} \tag{4.27}$$

The assumption of absorption leads to

$$1-f-b>0. \tag{4.28}$$

Standard reasoning then gives

$$\begin{aligned} \frac{d}{dx}l_r(a,x) &= \sigma\{(1-f-b)+(f-1)l_r(a,x) \\ &\quad +(1-f-b)r_r(a,x)+br_r(a,x)l_r(a,x)\}, \end{aligned} \tag{4.29}$$

$$l_r(a,a)=0.$$

The extension of this reasoning to the matrix case of transport theory is relatively easy. Starting with Eq. (1.32) we define a dissipation matrix L_r by setting

$$l_{r,ij}(a,x) = \text{average number of particles lost } \textit{within} \text{ the system each second while in state } i \text{ due to an input at the right side of the system of one particle each second in state } j. \tag{4.30}$$

$$L_r(a,x) = (l_{r,ij}(a,x)).$$

Then (see Problem 6)

$$\frac{d}{dx}L_r(a,x) = H + L_r(a,x)F + HR_r(a,x) + L_r(a,x)BR_r(a,x), \qquad L_r(a,a) = 0. \tag{4.31}$$

Here H is a diagonal matrix whose jth element is

$$h_{jj} = \sigma_j\left\{1 - \sum_{i=1}^{n}(f_{ij} + b_{ij})\right\}. \tag{4.32}$$

Clearly all elements of this diagonal matrix are nonnegative. We may as well assume that at least one is actually positive, for otherwise we have the conservative case already dealt with.

The final extension is to matrix equations of the form (4.3). A rather obvious, but somewhat tedious, set of analogies yields

$$\frac{d}{dx}L_r(a,x) = G + L_r(a,x)D + G'R_r(a,x) + L_r(a,x)CR_r(a,x), \tag{4.33}$$

$$L_r(a,a) = 0,$$

where now G and G' are diagonal matrices with elements

$$g_{jj} = -\sum_{i=1}^{n}(b_{ij} + d_{ij}), \tag{4.34a}$$

$$g'_{jj} = -\sum_{i=1}^{n}(a_{ij} + c_{ij}). \tag{4.34b}$$

If one now asks for the conditions on the matrices $A, B, \ldots,$ which are analogous to those which prevail in the actual physical nonconservative case, he finds (see Problem 9)

$$a_{ij} \geqslant 0 \quad \text{and} \quad d_{ij} \geqslant 0 \quad \text{for } i \neq j; \tag{4.35a}$$

$$b_{ij} \geqslant 0 \quad \text{and} \quad c_{ij} \geqslant 0 \quad \text{for all } i \text{ and } j; \tag{4.35b}$$

$$g_{jj} \geqslant 0 \quad \text{and} \quad g'_{jj} \geqslant 0 \tag{4.35c}$$

with at least one inequality holding in Eq. (4.35c). We shall refer to a system satisfying these conditions as a "generalized" nonconservative system.

7. THE EXISTENCE PROOF

We are finally in a position to sketch out the promised proof of existence of the solution everywhere to the generalized nonconservative problem. We first observe that in view of the conditions imposed on the coefficient matrices, we have

$$M(G' + A + C) = 0 \quad \text{and} \quad M(G + B + D) = 0. \tag{4.36}$$

Using the differential equations for R_r and T_r as in the previous section we readily ascertain that they have solutions over some interval $0 \leqslant x \leqslant x_0$, and that these solutions are again nonnegative. Moreover, it is equally readily seen from Eq. (4.33) that the same statements may be made concerning L_r. Now consider

$$\begin{aligned}
\frac{d}{dx} M(R_r + T_r + L_r) &= M(B + AR_r + R_rD + R_rCR_r) \\
&\quad + M(T_rD + T_rCR_r) + M(G + L_rD + G'R_r + L_rCR_r) \\
&= M[(B+G) + (R_r + T_r + L_r)D + (A + G')R_r + (R_r + T_r + L_r)CR_r] \\
&= M[(G + B + D) + (R_r + T_r + L_r - I)D \\
&\quad + (G' + A + C)R_r + (R_r + T_r + L_r - I)CR_r] \\
&= M[(R_r + T_r + L_r - I)(D + CR_r)].
\end{aligned} \tag{4.37}$$

Since at $x = a$, $R_r + T_r + L_r = I$, the unique solution to (4.37) is $M[R_r(a,x) + T_r(a,x) + L_r(a,x) - I] = 0$. This holds, of course, for $a \leqslant x \leqslant x_0$. But the nonnegativity of the matrices R_r, T_r, and L_r now implies that none of their elements exceeds unity in value. At this point precisely the same argument as that used at the end of the preceding section can be employed and the solution extended beyond x_0. This completes the proof.

Before making a formal statement of the theorem we have proved, let us observe again that the result just obtained is precisely what we should have expected physically. If we count all the particles emergent from a system and all the particles actually lost within the system, we come out with exactly the number of particles introduced into the system. On the other hand, *no* physical arguments whatsoever have been used in our proof. While it is true that the equation for the dissipation function L_r was vital and we were led to that equation by particle counting, we could just as well have introduced Eq. (4.33) completely abstractly with no reference to its origin. Physics has pointed the way, but the final result is independent of physical reasoning.

Let us now state formally what has been accomplished in this chapter.

THEOREM. Given the equation (4.3) subject to the conditions $u(a)=0$, $v(x)=v_x$. Let A, B,...,be constant matrices subject to the restrictions (4.35), G and G' being given by (4.34). Then the stated problem has a unique solution for all values of x, $a \leqslant x$.

COROLLARY. Under the hypotheses of the theorem the problem with $u(a)=u_a$ and $v(x)=v_x$ also has a unique solution for all x, $a \leqslant x$.

The corollary results from application of the functional equations of Chapter 3.

8. SUMMARY

In this chapter it has been demonstrated that the method of invariant imbedding can be used to obtain in a rigorous fashion information concerning the existence of solutions to certain classes of linear two-point boundary-value problems. Physical insight has played an important role, but the final results are independent of nonmathematical arguments.

Our presentation has suffered from the restriction that only problems with constant coefficients have been examined. This has been solely for convenience of exposition. We have chosen to leave the variable coefficient case to the problem section (Problems 2, 10–12).

Some rather strong conditions have been put on the coefficients in order that we could show existence of the solution for systems of arbitrary "length." While it is obviously true that one cannot completely abandon these hypotheses and still obtain such general results, it is very desirable to lessen these restrictions and then obtain information—in the form of estimates or bounds—on the length of the system for which a solution exists. Problems of this kind are quite difficult and have not yet been investigated despite their obvious importance.

PROBLEMS

1. Devise a model of a two-state physical system for which there are two distinct critical lengths depending upon whether the source particle is in state one or state two. Generalize to the n-state case.

2. Analyze the scalar model in which the parameters a, b, etc., are functions of z. Show that the results of this chapter extend to this case without difficulty.

3. Obtain Eq. (4.15) using the functions $U_1(z)$ and $V_1(z)$.

4. Prove, using classical methods, that the sequence generated by (4.22) converges to the unique solution of (4.21).

5. Carry through the arguments needed to prove the nonnegativity of T_r (see Section 5).

6. Give a detailed derivation of (4.27). Extend this to obtain (4.31). Consider also the situation in which F and B depend upon z.

7. Verify (4.33), first in the scalar, then in the matrix case. Generalize as in Problem 6.

8. Verify the assertion concerning L_r which occurs in the text just preceding (4.37).

9. Show that Eqs. (4.35) do indeed provide a mathematical generalization of the physically nonconservative case.

10. Check all the reasoning for the conservative case and confirm that the only place where *essential* use of the constancy of matrices has been made is in (4.21).

It is the form of (4.21) that allows direct use of the Lemma and leads to the conclusion that R_r is a nonnegative matrix. To avoid use of this approach we suggest the following attack: Consider the system

$$\frac{dz_i(t)}{dt} = \tilde{h}_{ii}(t)z_i(t) + \sum_{\substack{j=1 \\ j\neq i}}^{n} \tilde{h}_{ij}(t)z_j(t), \qquad t \geqslant 0,$$

where $\tilde{h}_{ij}(t) \geqslant 0$, $i \neq j$, $i,j = 1, 2, 3, \ldots, n$, and the $\tilde{h}$'s are all continuous functions. Let the initial conditions $z_i(0) = a_i$, $a_i \geqslant 0$ be imposed. Prove that the solution functions $z_i(t)$ are all nonnegative.

Generalize this result to the case

$$\frac{dz_i(t)}{dt} = \tilde{h}_{ii}(t)z_i(t) + \sum_{\substack{j=1 \\ j\neq i}}^{n} \tilde{h}_{ij}(t)z_j(t) + f_i(t),$$

where the f_i are nonnegative continuous functions.

Finally use this result to show that even when the matrices A, $B, \ldots,$in the text are nonconstant but continuous, the corresponding R_r matrix is nonnegative. Do this by defining a suitable sequence of matrices $R_{r,n}$ recursively.

Generalize the above results to include the case of matrices A, $B, \ldots,$which are only piecewise continuous.

11. Extend the reasoning of the text for the nonconservative case to cover nonconstant matrices A, $B,\ldots,$as in Problem 10.

12. The final results of Problems 10 and 11 may also be obtained as follows. The text proofs are rather easily extended to the case of matrices A, $B,\ldots,$which are *step* functions. Any piecewise continuous function may be uniformly approximated by such functions. For the case of A, $B,\ldots,$merely piecewise continuous consider a sequence of problems whose matrix coefficients are step functions which approximate the piecewise continuous A, $B,\ldots$. Show that a subsequence of the solutions of these problems converges to the solution of the given problem.

REFERENCES

1. R. Bellman, R. Kalaba, and G. M. Wing, "Dissipation Functions and Invariant Imbedding," *Proc. Nat. Acad. Sci. USA* **46**, 1960, 1145.
2. R. Bellman, K. L. Cooke, R. Kalaba, and G. M. Wing, "Existence and Uniqueness Theorems in Invariant Imbedding. I: Conservation Principles," *J. Math. Anal. Appl.* **10**, 1965, 234–244.
3. E. Coddington and N. Levinson, *Theory of Ordinary Differential Equations*, McGraw-Hill, New York, 1955.

5

RANDOM WALK

1. INTRODUCTION

There is a very close connection—both physical and mathematical in nature—between transport problems and random walk problems. It is therefore not at all surprising that the method of invariant imbedding should be applicable to random walk phenomena. We shall discuss matters of this kind in the present chapter, pointing out situations in which the imbedding method may be conceptually or computationally superior to the more classical approaches. Since many of the basic ideas are relatively straightforward generalizations of what we have been doing in earlier chapters results will often be given rather sketchily, with the details left to the problems.

In what follows we shall in some instances be introducing new types of imbeddings. It is important for the reader to recognize that the basic idea we are discussing is really quite general and that any particular problem may lend itself to many different imbedding approaches. Perhaps these different approaches are most easily understood in the context of random walk, since the discrete aspects of the problem make the required manipulations rather more transparent than is frequently the case in continuous models.

2. A ONE-DIMENSIONAL RANDOM WALK PROCESS

Let the points $[a,a+1,\ldots,k-1,k,k+1,\ldots,b-1,b]$ be "lattice points" arrayed on a line as indicated in Figure 5.1. We consider a moving "particle" located at the point k. The particle will either jump one step to the left or one step to the right with an assigned probability. We may assume that such jumps occur at unit time intervals, although frequently time will not enter explicitly into the analysis. Thus

$$p(k) = \text{probability that the particle, located at } k, \text{ will jump one unit step to the left;} \tag{5.1a}$$

$$q(k) = \text{probability that the particle, located at } k, \text{ will jump one unit step to the right.} \tag{5.1b}$$

We suppose

$$p(k) + q(k) = 1. \tag{5.2}$$

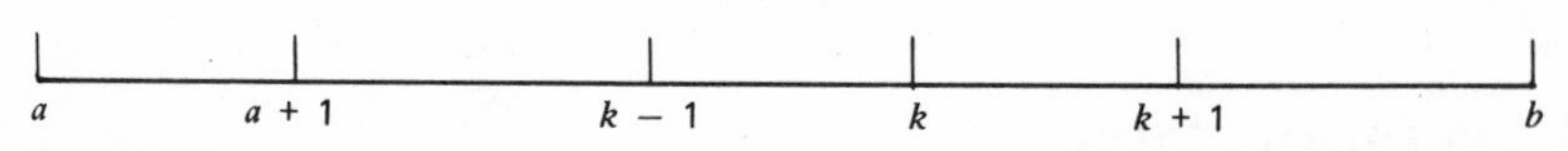

Figure 5.1.

That is, the particle must move from the point k. This is equivalent in the transport analogy to the assumption of no absorption. Note that p and q are assumed to depend only upon the location on the lattice, and not on the time. For convenience we require

$$0 < p(k) < 1. \tag{5.3}$$

This condition may be relaxed, but the assumption makes many of our analyses more straightforward, eliminating numerous special cases.

Next, let us suppose that the "random walk" described terminates if the particle reaches either a or b. In the customary terminology these end points are "absorbing barriers." Notice that there is complete analogy with the transport assumption that a moving particle is lost when it reaches an extremity of the system.

3. A CLASSICAL FORMULATION

We introduce the function

$$\pi(k) = \text{probability that a particle, starting at } k, \text{ reaches } a \text{ before it reaches } b. \tag{5.4}$$

This function is defined for $k = a+1, a+2, \ldots, b-1$, and we set

$$\pi(a) = 1, \qquad \pi(b) = 0. \tag{5.5}$$

A simple classical analysis yields the equation

$$\pi(k)=p(k)\pi(k-1)+q(k)\pi(k+1), \qquad a<k<b. \tag{5.6}$$

This is a linear difference equation subject to the two-point boundary conditions given by (5.5). Thus, the system is quite analogous to the two-point differential equation problems we have been encountering.

It is interesting and important to note that this problem has a unique solution. To see this, set $\pi'(b)=0$ and $\pi'(b-1)=c$, where c is positive but otherwise arbitrary. Let $\pi'(k)$ be obtained from (5.6) by computing "backward" from $b-1$. Thus, in particular,

$$\pi'(b-2)=\frac{c}{p(b-1)}>c=\pi'(b-1). \tag{5.7}$$

An easy induction argument now confirms that for all k, $k=a, a+1,\ldots, b-2$

$$\pi'(k)>\pi'(k+1)>0. \tag{5.8}$$

It is clear that in general $\pi'(a)\neq 1$, and thus the boundary condition of (5.5) is violated. However, since $\pi'(a)>0$ we may define

$$\pi(k)=\frac{\pi'(k)}{\pi'(a)} \tag{5.9}$$

and readily verify that we have found a solution to the problem. Incidentally, that the solution is positive for $k<b$ is clear.

The uniqueness is also quite trivial. Suppose the system (5.5)–(5.6) has more than one solution. Since the system is linear, the difference of any two solutions to (5.6) is a solution to (5.6). If $\eta(k)$ denotes this difference then

$$\eta(k)=p(k)\eta(k-1)+q(k)\eta(k+1). \tag{5.10}$$

Moreover,

$$\eta(a)=\eta(b)=0. \tag{5.10a}$$

Now suppose that for some $k=k'$ we have $\eta(k')\neq 0$. It will do no harm to suppose that indeed k' is the first lattice point to the left of b where this happens. Assume $\eta(k')>0$. Then an analysis identical to that used in showing (5.8) confirms that

$$\eta(k)>\eta(k+1)>0 \tag{5.11}$$

for all k, $k=a, a+1,\ldots,k'-1$. Thus $\eta(a)>0$ in violation of (5.10a).

The assumption that $\eta(k')<0$ is as readily used.

Clearly this result is the analog of the one found in the previous chapter in the (generalized) case of no absorption. It is interesting to note that in this random walk problem absorption can readily be included by an "ingenious device" and that the above reasoning therefore extends easily to the more general case (see Problem 1).

4. AN INVARIANT IMBEDDING FORMULATION

Still considering the model of Section 2 we now introduce the function

$$\hat{\pi}(a,k) = \text{the probability that a particle, starting at } k\text{, will reach } a \text{ before it reaches } b. \tag{5.12}$$

Evidently this is precisely the function defined by (5.4) except that we have seen fit to exhibit the role of a in an explicit fashion. Moreover, we shall now consider a as a discrete variable capable of being any one of the lattice points to the left of b. (Of course, a is also to the left of k or coincident with it.) As before

$$\hat{\pi}(a,a)=1, \qquad \hat{\pi}(a,b)=0. \tag{5.13}$$

To obtain a new equation for $\hat{\pi}(a,k)$ we use the obvious fact that a particle reaching a must first have reached $a+1$ at some previous time, since only a unit step in either direction is possible. This simple observation immediately yields

$$\hat{\pi}(a,k)=\hat{\pi}(a+1,k)\hat{\pi}(a,a+1). \tag{5.14}$$

Iterating this result gives

$$\hat{\pi}(a,k)=\hat{\pi}(a,a+1)\hat{\pi}(a+1,a+2)\cdots\hat{\pi}(k-1,k). \tag{5.15}$$

Thus the solution of the original problem is reduced to a calculation of $\hat{\pi}(a,a+1)$ for *any* a.

To find $\hat{\pi}(a,a+1)$ we note that Eq. (5.6) written in the new notation gives

$$\begin{aligned}\hat{\pi}(a,a+1)&=p(a+1)\hat{\pi}(a,a)+q(a+1)\hat{\pi}(a,a+2)\\&=p(a+1)+q(a+1)\hat{\pi}(a,a+2).\end{aligned} \tag{5.16}$$

Equations (5.14) and (5.16) now yield

$$\hat{\pi}(a,a+1)=\frac{p(a+1)}{1-q(a+1)\hat{\pi}(a+1,a+2)}. \tag{5.17}$$

Observe that the denominator of this expression cannot vanish since $q(a+1)<1$ and $\hat{\pi}(a+1,a+2)\leqslant 1$. Iteration produces a continued fraction

$$\hat{\pi}(a,a+1)=\cfrac{p(a+1)}{1-\cfrac{q(a+1)p(a+2)}{1-\cfrac{q(a+2)p(a+3)}{1-\cdots}}} \tag{5.18}$$

This expansion terminates because of the condition $\hat{\pi}(b-1,b)=0$.

5. SOME REMARKS CONCERNING SECTION 4

It is interesting to note that existence and uniqueness of the solution to the problem we have posed is immediate when the invariant imbedding approach is used, for we have actually constructed that solution. This is in distinct contrast to the equivalent questions that arise when the classical approach is employed. It may be argued, of course, that it is still necessary to prove that the two formulations are equivalent, and this is indeed true. However, there is a human tendency to "suspect" the new approach while accepting the older one without question. Conceptually, the invariant imbedding attach is no more subtle or difficult than is the standard method—it is only "different." The onus for showing equivalence lies as heavily on the classical equations as on the invariant imbedding equations. We cite this, not as a reason for overlooking the entire problem, but in an effort to put the question in some perspective.

Although the continued fraction expression for $\hat{\pi}(a,a+1)$ has a certain elegance, it is not immediately valuable from a computational viewpoint. In practice, one would probably be well advised to use the fact that since $\hat{\pi}(b-1,b)=0$ Eq. (5.17) yields

$$\pi(b-2,b-1)=p(b-1),$$

$$\hat{\pi}(b-3,b-2)=\frac{p(b-2)}{1-q(b-2)\hat{\pi}(b-2,b-1)}, \quad \text{etc.} \tag{5.19}$$

Thus one actually computes "backwards" to the value of a of interest.

Finally, we call attention to the fact that our results may be used to obtain estimates of the $\hat{\pi}$'s. These may be seen from either (5.17) or (5.18). Thus, for example,

$$\hat{\pi}(a,a+1) > \frac{p(a+1)}{1-\dfrac{q(a+1)p(a+2)}{1-q(a+2)p(a+3)}} > \frac{p(a+1)}{1-q(a+1)p(a+2)} > p(a+1). \tag{5.20}$$

These bounds obviously produce estimates for $\hat{\pi}(a,k)$. Such results are not nearly so obvious when the approach of Section 3 is employed.

6. SKETCH OF ANOTHER APPROACH

In this section we shall outline very briefly a scheme for an imbedding very similar to that used in the early chapters of this book. Because of this similarity we shall not go into very much detail, leaving much of the analysis to the problems. Moreover, the ideas have many points of contact with those used in a very natural way in Chapter 10 of this volume. The interested reader may wish to consult the material of that chapter. This may be done without reading the intervening material.

Let us write

$$u_k = \pi(k). \tag{5.21}$$

Rather direct manipulations may be used to put Eq. (5.6) in the form

$$u_{k+1} - 2u_k + u_{k-1} = A_k u_k + B_k u_{k-1}. \tag{5.22}$$

Define

$$\Delta u_k = u_{k+1} - u_k = v_k \tag{5.23}$$

so that (5.22) may be rewritten

$$v_k - v_{k-1} = A_k(u_{k-1} + v_{k-1}) + B_k u_{k-1}. \tag{5.24}$$

With obvious notational changes we then have

$$\Delta u_k = v_k,$$

$$-\Delta v_k = C_k u_k + D_k v_k. \tag{5.25}$$

Equation (5.25) is a system reminiscent of the systems encountered in early chapters, except, of course, that our new equations are difference equations rather than differential equations. It is also evident that it is really not particularly relevant that our system arose from a random walk problem; any second order difference equation in u_k could have been similarly treated. Moreover, the decomposition into u and v equations is by no means unique. This may be accomplished in many different ways. Finally, it is clear that the conditions on u_k at the two extremities of the system (analogous to $\pi(a)=1$ and $\pi(b)=0$) are not directly analogous to the differential equation case. Much more to be expected would be a condition on u at one end and upon v at the other.

However, it is completely reasonable to expect that a theory like that of Chapters 1–3 can be developed for equations of the form (5.25), including r and t functions, functional equations, etc., and that out of this theory one can obtain all of the necessary tools to handle the original random walk problem and its generalizations. This program can indeed be carried out. As noted earlier, the parallels with the work we have done are really so close that it seems appropriate to leave most of the details to the problems.

7. EXPECTED SOJOURN

Up to this point we have been suppressing considerations of time in the random walk problem, although it has been mentioned that it is convenient to consider a particle "jump" to take place every unit time. Let us now introduce the function

$$\tau(a,k)=\text{the expected time taken by a particle starting at } k \text{ to reach } a, \text{ assuming } a \text{ is reached before } b. \tag{5.26}$$

Recalling that we are now dealing with expectations rather than probabilities, we find rather easily that

$$\tau(a,k)=\tau(a+1,k)+\tau(a,a+1). \tag{5.27}$$

An equation for $\tau(a,a+1)$ is a trifle more difficult to obtain. If the particle at $a+1$ moves to the left then precisely one time unit expires before the event is completed. However, if the particle moves to the right, one time unit has been used up and the particle must now still get from $a+2$ to a. These two observations lead to

$$\tau(a,a+1)=p(a+1)+q(a+1)[\tau(a,a+2)+1]. \tag{5.28}$$

Now using (5.27) with $k=a+2$ we obtain

$$\tau(a,a+1)=\frac{1}{p(a+1)}+\frac{q(a+1)}{p(a+1)}\tau(a+1,a+2). \tag{5.29}$$

An obvious iteration produces

$$\tau(a,a+1)=\frac{1}{p(a+1)}+\frac{q(a+1)}{p(a+2)p(a+1)}+\frac{q(a+1)q(a+2)}{p(a+3)p(a+2)p(a+1)}\cdots. \tag{5.30}$$

Elementary reasoning yields

$$\tau(b-2,b-1)=1/p(b-1). \tag{5.31}$$

Moreover, the relationship (5.31) suggests that a practical method of computing $\tau(a,a+1)$ is to procede "backwards" from b. This is analogous to our findings with respect to the computation of the function $\hat{\pi}$ in Section 4.

It will be noted that in this brief study of expected sojourn we have gone immediately to an invariant imbedding type of approach. A more classical attack is also possible, a matter we leave to the problem section.

8. A "MANY-STATE" CASE—INVARIANT IMBEDDING APPROACH

We now turn to a more complex random walk process, analogous to some of the transport phenomena studied in the early chapters of this book. We assume that the particle performing the walk can exist in any one of a finite number of "states," and that in moving from one lattice point to another its state may change. In the transport problems discussed earlier we also introduced the concept of "state," noting there that energy is perhaps the state of most frequent physical interest. In random walk problems our imaginations can carry us farther afield. It is rather amusing, for example, to imagine a particle that may actually change color as it moves about on the line. Regardless of what the concept of state means in a particular situation, we can always employ an index i, $i=1,2,\dots,N$, to indicate what particular state the particle is in. Let us suppose that

$p_{ij}(k)=$probability that the particle in state j

and at lattice point k moves to the left

one step and arrives at lattice point

$(k-1)$ in state i. (5.32)

The probability $q_{ij}(k)$ for movement to the right is defined similarly. We also require

$$\sum_{i=1}^{N} [p_{ij}(k) + q_{ij}(k)] = 1, \qquad j = 1, 2, \ldots, N. \tag{5.33}$$

Equation (5.33) states that the particle must jump either to the right or to the left and arrive in some state. In other words, there is no absorption.

Now set

$$\pi_{ij}(a,k) = \text{probability that a particle starting at lattice point } k \text{ in state } j \text{ will reach } a \text{ in state } i \text{ before it reaches } b \text{ in any state.} \tag{5.34}$$

A bit of thought and calculation reveals that

$$\pi_{ij}(a,k) = \sum_m \pi_{mj}(a+1,k)\pi_{im}(a,a+1). \tag{5.35}$$

The form of (5.35) suggests that we introduce the matrix $\Pi(a,k)$ whose elements are the π_{ij}. Then

$$\Pi(a,k) = \Pi(a,a+1)\Pi(a+1,k), \tag{5.36}$$

an equation reminiscent of (5.14). Again

$$\pi_{ij}(a,a+1) = p_{ij}(a+1) + \sum_m q_{mj}(a+1)\pi_{im}(a,a+2), \tag{5.37}$$

or, in matrix form,

$$\Pi(a,a+1) = P(a+1) + \Pi(a,a+2)Q(a+1), \tag{5.38}$$

where the meaning of P and Q is clear.

If we set $k=2$ in (5.36) and use the result in (5.38) we obtain formally

$$\Pi(a,a+1) = P(a+1)[I - \Pi(a+1,a+2)Q(a+1)]^{-1}. \tag{5.39}$$

The existence of the inverse must be demonstrated, of course. From this point on the discussion and analysis parallels that of Section 4 and we leave details to the problem section.

9. TIME DEPENDENT PROCESSES—CLASSICAL APPROACH

While mention has been made of the fact that we may consider the particle performing the random walk to make one jump in each unit of time, time has been explicitly included only in our analysis of expected sojourn. We now define

$$\pi(a,k,t) = \text{the probability that the particle starting at } k \text{ will reach the point } a \text{ in exactly time } t \text{ without ever arriving at the point } b. \tag{5.40}$$

Note that t assumes only integer values. Observe, too, that we refrain from introducing awkward symbols such as $\hat{\pi}$ and do not hesitate to reuse in a different context the symbol employed with a slightly different meaning in previous sections.

First, let us analyze this problem in the classical manner. In other words, we perform a "classical imbedding." Quite clearly

$$\pi(a,k,t) = p(k)\pi(a,k-1,t-1) + q(k)\pi(a,k+1,t-1) \tag{5.41}$$

provided $a < k < b$ and $t \geqslant 1$. We also have the initial condition

$$\pi(a,k,0) = \delta_{ak}, \tag{5.42}$$

where δ is the Kronecker delta. The boundary conditions are

$$\pi(a,a,t) = 0, \qquad \pi(a,b,t) = 0 \qquad \text{for } t \geqslant 1. \tag{5.43}$$

A standard procedure for solving problems of this type is to introduce a generating function. In this case we define

$$g(a,k,r) = \sum_{t=0}^{\infty} \pi(a,k,t)r^t. \tag{5.44}$$

This series makes sense, and, indeed, defines an analytic function, for $|r| < 1$. Using (5.41) and manipulating a bit, we obtain,

$$\begin{aligned} g(a,k,r) &= \sum_{t=1}^{\infty} r^t[p(k)\pi(a,k-1,t-1) + q(k)\pi(a,k+1,t-1)] \\ &\quad + \pi(a,k,0) \\ &= rp(k)g(a,k-1,r) + rq(k)g(a,k+1,r), \end{aligned} \tag{5.45}$$

for $a<k<b$ with

$$g(a,a,r)=1, \qquad g(a,b,r)=0. \tag{5.46}$$

Thus, the function g satisfies a difference equation in k very similar to the one satisfied by the function π of Section 3. We leave the detailed analysis as a problem, mentioning only that once $g(a,k,r)$ has been found explicitly it may be expanded in a power series in r and the coefficients of that series are precisely the desired probabilities $\pi(a,k,t)$.

10. TIME-DEPENDENT PROCESSES—INVARIANT IMBEDDING APPROACH

By now the reader should have little difficulty in verifying that the invariant imbedding method applied to the problem described in the previous section yields

$$\pi(a,k,t)=\sum_{s=0}^{t}\pi(a+1,k,s)\pi(a,a+1,t-s), \qquad k\geqslant a+1. \tag{5.47}$$

This "convolution-like" expression suggests that we again use the generating function $g(a,k,r)$. The result is

$$g(a,k,r)=g(a,a+1,r)g(a+1,k,r). \tag{5.48}$$

From (5.41) through (5.43),

$$\begin{aligned}
&\pi(a,a+1,0)=0,\\
&\pi(a,a+1,1)=p(a+1),\\
&\pi(a,a+1,t)=q(a+1)\pi(a,a+2,t-1), \qquad t>1,
\end{aligned} \tag{5.49}$$

so that

$$g(a,a+1,r)=rp(a+1)+rq(a+1)g(a,a+2,r). \tag{5.50}$$

Using (5.48) with $k=a+2$ and (5.50) yields

$$g(a,a+1,r)=\frac{rp(a+1)}{1-rq(a+1)g(a+1,a+2,r)}. \tag{5.51}$$

This equation has the same relationship to (5.17) of Section 4 as do (5.45) and (5.46) to (5.5) and (5.6). The overall analysis of (5.51) parallels that of (5.17) and we again leave the details to the problems.

11. A MULTISTEP PROCESS—CLASSICAL APPROACH

Let us now consider a somewhat more general problem than the basic one we have been dealing with. Suppose the moving particle is capable of taking a jump of either one unit or two units in either direction. Specifically, let

$$\begin{aligned} p_i(k) &= \text{the probability of the particle moving } i \\ &\quad \text{steps to the left when it is at } k,\ i=1,2; \\ q_i(k) &= \text{the probability of the particle moving } i \\ &\quad \text{steps to the right when it is at } k,\ i=1,2. \end{aligned} \tag{5.52}$$

Suppose the lattice points are still arrayed from a to b. It is clear that if the particle is at $b-1$ it may make a jump of two units to the right and miss the "absorbing barrier" b completely. We shall still assume that the particle has disappeared from the system; effectively it has still been absorbed at b. We must also recognize that a particle at $a+1$ may make a jump of two units to the left and miss a completely. For convenience we can suppose this particle to have landed at $a-1$. Thus we introduce two functions:

$$\begin{aligned} \pi(a,k) &= \text{the probability that the particle starting} \\ &\quad \text{at } k \text{ will actually reach } a \text{ before it} \\ &\quad \text{disappears at } b; \\ \hat{\pi}(a-1,k) &= \text{the probability that the particle starting} \\ &\quad \text{at } k \text{ will actually reach } a-1 \\ &\quad \text{(without reaching } a\text{) before it disappears} \\ &\quad \text{at } b. \end{aligned} \tag{5.53}$$

In the above we assume $a<k<b$. We set

$$\begin{aligned} &\pi(a,b) = \hat{\pi}(a-1,b) = 0, \\ &\pi(a,a) = 1, \qquad \hat{\pi}(a-1,a) = 0. \end{aligned} \tag{5.54}$$

Now the standard procedures yield, for $k=a+2,a+3,\dots,b-2$,

$$\pi(a,k)=p_1(k)\pi(a,k-1)+p_2(k)\pi(a,k-2)+q_1(k)\pi(a,k+1)$$
$$+q_2(k)\pi(a,k+2),$$
$$\hat{\pi}(a-1,k)=p_1(k)\hat{\pi}(a-1,k-1)+p_2(k)\hat{\pi}(a-1,k-2)$$
$$+q_1(k)\hat{\pi}(a-1,k+1)+q_2(k)\hat{\pi}(a-1,k+2). \qquad (5.55)$$

The cases $k=a+1$ and $k=b-1$ are somewhat special. We write the results for $k=a+1$, leaving the others as Problem 10.

$$\pi(a,a+1)=p_1(a+1)+q_1(a+1)\pi(a,a+2)+q_2(a+1)\pi(a,a+3),$$
$$\hat{\pi}(a-1,a+1)=p_2(a+1)+q_1(a+1)\hat{\pi}(a-1,a+2)+q_2(a+1)\hat{\pi}(a-1,a+3).$$
$$(5.56)$$

The solution of this set of difference equations is obtainable by fairly standard techniques and we do not pursue the matter. Of interest is the fact that the problem is once again of the two-point boundary value type. This often provides a threatening situation when numerical results are desired.

12. A MULTISTEP PROCESS—INVARIANT IMBEDDING APPROACH

The model just described may be handled quite elegantly by invariant imbedding. The basic equations are

$$\pi(a,k)=\pi(a,a+2)\pi(a+2,k)+\pi(a,a+1)\hat{\pi}(a+1,k),$$
$$\hat{\pi}(a-1,k)=\hat{\pi}(a-1,a+2)\pi(a+2,k)+\hat{\pi}(a-1,a+1)\hat{\pi}(a+1,k), \qquad (5.57)$$

under the restriction $k \geqslant a+3$. While other forms may be developed, (5.57) is very convenient because it may be written as

$$\Pi(a,k)=\Pi(a,a+1)\Pi(a+2,k), \qquad (5.58)$$

where

$$\Pi(a,k)=\begin{pmatrix} \pi(a,k+1) & \pi(a,k) \\ \hat{\pi}(a-1,k+1) & \hat{\pi}(a-1,k) \end{pmatrix}. \qquad (5.59)$$

For (5.58) to be useful it is necessary to know $\Pi(a,a+1)$. For this purpose we turn to (5.55) and (5.56) with $k=a+2$. Some manipulation reveals that

$$\pi(a,a+1)=\alpha_1+\beta_1\pi(a,a+3)+\gamma_1\pi(a,a+4),$$

$$\pi(a,a+2)=\alpha_2+\beta_2\pi(a,a+3)+\gamma_2\pi(a,a+4), \tag{5.60}$$

where the expressions α_i, β_i, γ_i depend on the values of the p and q functions at $a+1$ and $a+2$. We leave their precise values to the problem section. Similarly

$$\hat{\pi}(a,a+1)=\hat{\alpha}_1+\beta_1\hat{\pi}(a,a+3)+\gamma_1\hat{\pi}(a,a+4),$$

$$\hat{\pi}(a,a+2)=\hat{\alpha}_2+\beta_2\hat{\pi}(a,a+3)+\gamma_2\hat{\pi}(a,a+4). \tag{5.61}$$

It is important to notice that the coefficients β_1, β_2, γ_1, and γ_2 are precisely the same as in (5.60). Hence we may write

$$\Pi(a,a+1)=P(a)+\Pi(a,a+3)Q(a),$$

$$P(a)=\begin{pmatrix}\alpha_2 & \alpha_1\\ \hat{\alpha}_2 & \hat{\alpha}_1\end{pmatrix}, \tag{5.62}$$

$$Q(a)=\begin{pmatrix}\gamma_2 & \gamma_1\\ \beta_2 & \beta_1\end{pmatrix}.$$

Using (5.62) in (5.58) with $k=a+3$ yields formally

$$\Pi(a,a+1)=P(a)[I-\Pi(a+2,a+3)Q(a)]^{-1}. \tag{5.63}$$

We leave the verification that the indicated inverse actually exists as a problem.

The similarity between Eqs. (5.63) and (5.17) is quite striking, suggesting the possibility of extension to problems in which the moving particle can make $1,2,\ldots,$ or m steps to either right or left with assigned probabilities. We also leave this matter to the reader. Here we merely note that just as in Section 4 the formula for $\Pi(a,a+1)$ is probably most easily applied by working "backwards" from the right, noting that (see Problem 13)

$$\Pi(b-2,b-1)=\begin{pmatrix}0 & p_1(b-1)\\ 0 & p_2(b-1)\end{pmatrix}. \tag{5.64}$$

13. SOME REMARKS ON AN EXTENSION TO A CONTINUOUS CASE

We now consider a "continuous" random walk problem in which the moving particle, still confined to the line, can move an arbitrary distance to right or to left at each jump. Let

$$k(t,s)\Delta+o(\Delta)=\text{the probability that a particle at } t \text{ will jump to a location between } s \text{ and } s+\Delta,\ \Delta>0. \tag{5.65}$$

Since we desire the particle definitely to land somewhere (and not simply disappear) we have the normalization

$$\int_{-\infty}^{\infty} k(t,s)\,ds=1. \tag{5.66}$$

Now define

$$\pi(t)=\text{the probability that a particle starting at } t,\ a\leqslant t\leqslant b, \text{ will escape to the left before it escapes to the right.} \tag{5.67}$$

Then standard reasoning yields the linear integral equation

$$\pi(t)=\int_{-\infty}^{a} k(t,s)\,ds+\int_{a}^{b} k(t,s)\pi(s)\,ds. \tag{5.68}$$

A case of particular importance is that when k is a difference kernel:

$$k(t,s)=g(|t-s|). \tag{5.69}$$

We might now try to formulate this random walk problem directly by invariant imbedding arguments. Instead of doing that we delay the entire analysis until Chapter 12, where the problem of resolving quite general linear Fredholm equations of the form

$$\varphi(t)=f(t)+\int_{a}^{b} g(|t-s|)\varphi(s)\,ds \tag{5.70}$$

by the invariant imbedding method, is considered. While certain restrictions must be put on f and g in that chapter, normalization such as (5.66) is not needed, nor is the nonnegativity which is implied in the present case by the definition of k and π.

14. SOME REMARKS ABOUT RANDOM WALK IN TWO DIMENSIONS

A classical extension of the easiest type of random walk problem involves the lattice points in the plane. For convenience, suppose a particle is located at a lattice point (h,k) and that that lattice point is itself in a region $\mathfrak{R}$, which for simplicity we may take to be a rectangle whose vertices are themselves lattice points. Let the probability of the particle moving to any of the four adjacent lattice points be $\frac{1}{4}$. Define

$$\pi(h,k) = \text{the probability that the particle starting at } (h,k) \text{ reaches a preassigned portion } B_1 \text{ of the boundary of } \mathfrak{R} \text{ before it reaches any other portion of the boundary, denoted } B_2. \tag{5.71}$$

If (h,k) is not itself a boundary point then

$$\pi(h,k) = \tfrac{1}{4}[\pi(h+1,k)+\pi(h-1,k)+\pi(h,k+1)+\pi(h,k-1)]. \tag{5.72}$$

If (h,k) is on B_1 then π has the value 1, while if it is on B_2 then π is zero.

This problem is especially interesting since it is an approximation to the continuous potential problem

$$\frac{\partial^2\varphi}{\partial x^2}(x,y)+\frac{\partial^2\varphi}{\partial y^2}(x,y)=0, \qquad (x,y)\in\mathfrak{R}, \tag{5.73}$$

$$\varphi(x,y)=1, \qquad (x,y)\in B_1; \qquad \varphi(x,y)=0, \qquad (x,y)\in B_2.$$

Various devices have been employed to use this approximate relationship as a means of getting estimates to the solution of potential problems, expecially when the region $\mathfrak{R}$ is of irregular shape instead of the simple rectangle we have assumed.

The importance of this type of consideration suggests the desirability of extending the invariant imbedding method to include such two-dimensional random walk problems. This has been done within certain limitations, and the device has proved quite successful. The most severe restriction is on the shape of the regions which may be conveniently considered. In the case of the rectangle the analysis is fairly straight forward, but even when the region is simply a union of rectangles certain difficulties of a practical nature arise. These may, however, be overcome, and useful numerical algorithms are obtained. We pursue the matter no farther, referring the reader to the problems and to the original papers and a recent book [1,2].

15. SUMMARY

In this chapter we have examined a rather wide variety of random walk problems. The discrete feature of the random walk has perhaps made the imbedding principle somewhat more transparent than has been the case in earlier chapters. Indeed, we have actually used several different imbedding procedures; even the classical method of handling such problems involves one type of imbedding. In many instances equations have been derived which provide easier computational methods for random walk problems than those classically found. However, no effort at obtaining or comparing actual numerical results has been made.

In a number of problems the analysis has not been carried through to completion. In some instances this is because the details are quite transparent and can be left to the reader. In others the amount of work involved is really extensive and it has not seemed reasonable to try to include more than a sketch in the present treatment. Indeed, the integral equation encountered in Section 13 will occupy all our effort in Chapter 12, while the analysis of higher dimensional random walk problems actually takes up a fair portion of a recent book [2]. Nevertheless, it is hoped that our brief discussion of these matters will convince the reader of the power of the imbedding technique and will lead him to make further investigations on his own.

PROBLEMS

1. Consider the model of Section 3, but now include the case of absorption at k:

$$y(k) = \text{probability that particle at } k \text{ does not move at all},$$

$$p(k)+q(k)+y(k)=1.$$

Derive the following equation for $\pi(k)$:

$$\pi(k)(1-y(k)) = p(k)\pi(k-1)+q(k)\pi(k+1)$$

with

$$\pi(k)=0 \qquad \text{if } y(k)=1.$$

Find appropriate boundary conditions.

Define, in the case $y(k)<1$,

$$\tilde{p}(k)=\frac{p(k)}{1-y(k)}, \qquad \tilde{q}(k)=\frac{q(k)}{1-y(k)}.$$

Hence reduce the problem to the one studied in the text (Section 3).

2. To what extent may be the "ingenious device" of Problem 1 be extended to other models considered in this chapter?

3. Suppose you were given only the invariant imbedding equations (5.15) and (5.17) and had no knowledge of the classical approach of Section 3. Could you obtain the ordinary random walk equations from these?

4. Determine the explicit form of the coefficients A_k, $B_k, \ldots$, in Section 6.

5.* Construct a theory of the kind alluded to in Section 6.

6. Attack the "expected sojourn" problem of Section 7 from the classical viewpoint instead of using the invariant imbedding approach.

7. Try to include the case of absorption in the many state problem.

8. Show that the inverse matrix indicated in (5.39) does indeed exist. Hence complete the analysis of the n-state problem.

9. Calculate the function $g(a,k,r)$ of Section 9 in complete detail when $p(k) = q(k) = \frac{1}{2}$ for all k. Expand it in a power series in r and thus obtain the coefficients $\pi(a,k,t)$. Complete the analysis of Section 10 to obtain the same result.

10. Consider the analog of (5.56) for the case $k = b-1$. Discuss the computational aspects of the solution of the entire problem.

11.* Generalize Section 11 and 12 to the case in which the particle may make jumps of $1,2,3,\ldots,n$ steps in either direction.

12. Calculate α_i, β_i, γ_i, and $\hat{\alpha}_i$ for Eqs. (5.60) and (5.61) explicitly. Confirm that the β_i and γ_i in the second equation are indeed the same as those in the first.

13. Prove that the indicated inverse in (5.63) exists. Also, establish the "initial condition" (5.64).

14.* Try to develop invariant imbedding equations for the problem posed in Section 14. Do this first for a rectangle and choose B_1 in a convenient fashion. Extend your reasoning to regions that are the unions of two rectangles, investigating any interesting difficulties that arise. Generalize to as arbitrary two-dimensional regions as you are able.

15. Consider the case where $q(k) = q^k$, $0 < q < 1$, and derive the equations for expected sojourn for a semi-infinite line, $a = 0$, $k = 1,2,3,\ldots$.

16. Obtain the uniqueness result of Section 3 by the following reasoning: Suppose there are two solutions to the problem. Let $d(k)$ be the difference of these solutions, and observe that $d(a) = d(b) = 0$. Let k_1 be a value of k for which $|d(k)|$ is largest. From Eq. (5.10) we find

$$|d(k_1)| \leqslant p(k_1)|d(k_1-1)| + q(k_1)|d(k_1+1)|.$$

By virtue of the choice of k_1 the equality must hold. (Why?) Thus $|d(k_1)| = |d(k_1-1)| = |d(k_1+1)|$. Repeating the argument we find eventually $|d(k_1)| = |d(a)| = 0$.

Actually this is an elementary application of the very basic maximum principle. Apply similar reasoning to the two-dimensional problems of Section 14 to obtain uniqueness results. Generalize in several different directions.

17.* Let u satisfy the equation $u=f+\lambda T(u)$ where f is a known function, λ is a parameter, and T is a linear transformation. Using the relation

$$T^n(u)=T^n(f)+\lambda T^{n+1}(u),$$

show that

$$\frac{T^n(u)}{T^{n+1}(u)}=\frac{T^n(f)}{T^{n+1}(f)}+\lambda\left[1-\frac{T^n(f)/T^{n+1}(f)}{T^{n+1}(u)/T^{n+2}(u)}\right].$$

Thus obtain a formal infinite continued fraction expansion for $u/T(u)$ and hence for u itself.

18.* Apply the results of Problem 17 to the equation $u=1+\lambda\int_0^t u\,dt_1$ to find a continued fraction expansion for e^t. (See R. Bellman and J. M. Richardson, "A New Formalism in Perturbation Theory Using Continued Fractions," *Proc. Nat. Acad. Sci., U.S.* **48**, 1962, 1913–1915.)

19.* Let u_n be the nth convergent of the infinite continued fraction in Problem 17. Show that

$$u_n=\sum_{i=0}^{n-1}\lambda^i T^i(f)+\frac{\lambda^n[T^n(f)]^2}{T^n(f)-\lambda T^{n+1}(f)}.$$

20. Consider the process of Section 13. Write Eq. (5.68) in the form

$$u(t)=f(t)+\int_a^b k(t,s)u(s)\,ds,$$

and suppose that $\int_{-\infty}^a k(t,s)\,ds=f(t)>0$. Show that this equation has a unique solution which may be written as a convergent series

$$u(t)=f(t)+\int_a^b k(t,s)f(s)\,ds+\int_a^b\int_a^b k(t,s')k(s',s)f(s)\,ds\,ds'$$

$$+\int_a^b\int_a^b\int_a^b k(t,s'')k(s'',s')k(s',s)f(s)\,ds\,ds'\,ds''+\cdots.$$

This is the Neumann series for the problem. Recall the physical origin of the equation, and try to interpret the meaning of each of the terms of the series.

21. Recall the properties of both f and k in Problem 20. Obtain the bounds

$$u(t)>f(t),\qquad u(t)>f(t)+\int_a^b k(t,s)f(s)\,ds,\qquad \text{etc.}$$

22. To obtain upper bounds for $u(t)$ as given in Problem 20 recall that $u(t)$ is really $\pi(t)$, the probability of escape to the left. Let $v(t)$ be the probability of escape to the right. Show that

$$v(t)=\int_b^\infty k(t,s)\,ds+\int_a^b k(t,s)v(s)\,ds.$$

Discuss the solution to this equation, including its Neumann series. Prove that $u(t)+v(t)=1$. Find lower bounds on $v(t)$, and hence upper bounds on $u(t)$.

23.* When can a linear equation of the form

$$u(t)=f(t)+\int_a^b k(t,s)u(s)\,ds$$

be associated with a random walk process? (See R. Bellman and R. Vasudevan, "Upper and Lower Bounds for the Solutions of Fredholm Integral Equations," *Utilitas Mathematica*, **4**, 1973, 221–229.)

24.* Consider the following stochastic process: An object born at time $t=0$ has a random lifetime with probability distribution given by $G(t)$. At the end of its life it is replaced by a random number of similar objects, p_r being the probability that the new objects are r in number, $r=0,1,2,\ldots$. The p_r are independent of the absolute time, the age of the object when it dies, and of the number of objects existing at the time.

Let $z(t)$ be the number of objects existing at time t. This is a random function referred to as an age-dependent branching process. Set

$$f_r(t)=\text{Prob}(z(t)=r),$$

$$g(s,t)=\sum_{r=0}^{\infty} f_r(t)s^r, \qquad |s|<1,$$

$$h(s)=\sum_{r=0}^{\infty} p_r s^r.$$

Show that

$$g(s,t)=s(1-G(t))+\int_0^t h(g(s,t-t'))\,dG(t'), \qquad t\geqslant 0.$$

25.* Let $v(t)=E(z(t))$, the expected number of objects at time t, as described in Problem 24. Show that

$$v(t)=1-G(t)+m\int_0^t v(t-t')\,dG(t'),$$

where $m=h'(1)$. Do this directly and also by differentiation of the nonlinear equation of the preceding problem.

26.* Consider Problems 24 and 25 when $dG(t)=ae^{-at}\,dt$, $a>0$. More generally, suppose dG is a sum of exponentials. (See R. Bellman and T. Harris, "On Age-dependent Binary Branching Processes," *Ann. Math.* **55**, 1952, 280–285.)

27. Let the integral equation

$$u(t)=f(t)+\int_a^b k(t,s)u(s)\,ds$$

be given. Here f and k may or may not have a specific physical meaning. An imbedding (due to Poincaré) may be achieved by introducing a parameter λ and considering the family

$$w(t)=f(t)+\lambda\int_a^b k(t,s)w(s)\,ds.$$

Construct a Neumann series for this equation. (See Problem 20.) It will be a power series in λ. Prove under reasonable conditions on f and k that it will converge for small λ. Now attempt to give a "pseudophysical" interpretation for each of the terms in the series.

28. Consider the infinite product $\prod_{k=1}^{\infty}(1+q^k)$. Show that this product is convergent for $|q|<1$. Examine the problem of evaluating the product numerically to a high degree of accuracy (say ten decimal places) for both small values of $|q|$ and values near unity.

29. Modify the product in Problem 28 to $\prod_{k=1}^{\infty}(1+xq^k)=f(x,q)$. (This imbedding technique is due to Euler.) Show that $f(x,q)$ is an analytic function of x for all x provided $|q|<1$. Next prove that $f(x,q)=(1+xq)f(xq,q)$. Use this functional equation to prove that f has the continued fraction expansion

$$f(x,q)=1+\cfrac{qx}{1-q+\cfrac{qx^2}{(1-q)(1-q^2)+\cdots}}.$$

Reconsider the question of numerical computation mentioned in the preceding problem.

30.* Obtain results similar to those of Problem 29 for the truncated product $\prod_{k=1}^{N}(1+xq^k)$.

REFERENCES

1. E. Angel, "Discrete Invariant Imbedding and Elliptic Boundary Value Problems over Irregular Regions," *J. Math. Anal. Appl.*, **23**, 1968, 471–484.
2. E. Angel and R. Bellman, *Dynamic Programming and Partial Differential Equations*, Academic, New York, 1972.

6

WAVE PROPAGATION

1. INTRODUCTION

In this chapter we shall study a class of problems which are much more frequently encountered in classical mathematical physics than are the transport questions we have investigated in the first three chapters. It was indicated in Chapter 1 that the earliest identifiable use of invariant imbedding was made by Stokes in the study of the reflection and transmission of light waves impinging upon a stack of glass plates [1]. We shall now investigate this type of phenomenon. Even at this point our methods are much more powerful than those available to Stokes, and we shall not only be able to reproduce his results but actually to go much further and study far more complicated problems than those that occupied his attention.

In keeping with our efforts to achieve variety, we shall approach the basic problem from several viewpoints. One is essentially physical and is quite analogous to the particle counting technique that has been employed in previous chapters. We might think of it as "wavelet counting." Historically, the first results using invariant imbedding on wave propagation problems were obtained in much this way. Other approaches will be much more analytical in nature. While this latter analysis in some ways obscures the physics of the matter, in other ways it actually makes the physical aspects more understandable. Indeed, the fundamental W. K. B. (Wentzel, Kramers, Brillouin) method of quantum mechanics, together with the less well known series of Bremmer, will be found as by-products of our results. The analytical approach possesses the advantage of reliability and ease of generalization in some directions; the "particle-wavelet counting" approach generalizes in other directions and provides the basis for some important approximation techniques because of the physical insight it provides.

We shall also discover in the course of our work a slight but important variant of the imbedding method we have thus far studied. In essence we shall find that the medium in which we imbed the "material" to be studied is rather more arbitrary than we might have previously imagined. The effect on the imbedding equations is quite striking.

2. THE CONCEPT OF A PLANE WAVE

We begin with the classical wave equation

$$\frac{d^2\psi(z)}{dz^2} + k^2(z)\psi(z) = 0, \qquad -\infty < z < \infty. \tag{6.1}$$

Although the physical waves represented by this equation may be of a large number of different kinds (waves in an infinitely long rope, quantum mechanical waves, acoustic waves, and so on) it may be helpful at times to fix ideas by thinking of light waves. Under the assumption that the quantity k in Eq. (6.1) is a positive constant, $k(z) \equiv k$, we note that two fundamental solutions are

$$\psi_l(z) = e^{-ikz}, \qquad \psi_r(z) = e^{+ikz}. \tag{6.2}$$

We consider ψ_l to represent a wave moving to the left while ψ_r represents a wave moving to the right. Obviously, the general solution to (6.1) is a linear combination of ψ_l and ψ_r, a simple superposition of right- and left-moving waves. Here the constants used in forming the linear combination may be complex. Thus if $c = ae^{i\theta}$, $a > 0$, then $c\psi_r = ae^{i(kz+\theta)}$ represents right-moving wave with *amplitude* a and *phase* θ. Clearly the phase is not a uniquely defined quantity, a matter which need not concern us here. For convenience in speaking we shall often not distinguish between the wave and the function that represents it; thus ψ_r will simply be referred to as a right-moving wave.

3. A TWO MEDIUM PROBLEM

It is instructive and extremely useful for the investigations that follow to begin with a very simple problem in wave theory. Let the half space $-\infty < z < x$ consist of a medium characterized by the constant $k = k_1 > 0$ in Eq. (6.1). The quantity k is usually referred to as the *wave number* of the medium. It is chosen to be positive. Let the half space $x \leqslant z < \infty$ have wave number k_2. Finally suppose a light wave travelling from $z = \infty$ to the left impinges on the interface $z = x$. We wish to solve the complete problem, determining the ψ function for all z.

We first notice that when the light strikes the interface some part of the wave may be reflected back into the right-hand medium while some may be transmitted into the medium on the left. Physically there is no mechanism for further reflection of the light once the wave is to the left of $z=x$. Thus in the left medium there is only a left-moving wave, while in the right-hand medium there are both left- and right-moving waves. We write

$$\psi(z)=e^{-ik_2(z-x)}+\rho(x)e^{ik_2(z-x)}, \qquad x<z<\infty;$$

$$\psi(z)=\tau(x)e^{-ik_1(z-x)}, \qquad -\infty<z<x. \tag{6.3}$$

The impinging wave has been chosen to have unit amplitude and the phase has been picked for convenience in further calculations. The quantities ρ and τ will be referred to as the *reflection* and *transmission* functions for the problem.

To reach an analytic solution we must impose additional conditions at the interface $z=x$. Physical considerations lead to the requirements

$$\psi(x+0)=\psi(x-0), \qquad \psi'(x+0)=\psi'(x-0). \tag{6.4}$$

Elementary calculations now yield

$$\rho(x)=\frac{k_2-k_1}{k_1+k_2},$$

$$\tau(x)=\frac{2k_2}{k_1+k_2}. \tag{6.5}$$

Two remarks are appropriate. First, the assumed amplitude and phase of the wave moving to the left from $z=\infty$ are only a matter of convenience. Had the wave been $se^{-ik_2(z-x)}$, where s is any real or complex constant, then the reflected and transmitted wave could be obtained by simply multiplying ρ and τ by s. This is an immediate consequence of the linearity of the defining equation.

Second, the physical phenomenon of *specular* reflection has been encountered. There is a right moving wave simply because the two media have different wave numbers. This "mirror-like" behavior was not found in the problems we studied in simple transport theory. It is further illuminated by investigating the three medium case, a matter we leave to the problems. (See Problem 1.)

4. A MULTIMEDIUM PROBLEM

We now pose a much more interesting physical problem. Suppose the two media of the previous section are replaced by $N+2$ media with interfaces at $x_0<x_1<x_2<\cdots<x_N$. The space to the left of x_0 has wave number k_0,

that to the right of x_N has number k_{N+1}, and the medium in $x_{j-1} \leqslant z < x_j$ is characterized by wave number k_j. A "unit" wave $e^{-ik_{N+1}(z-x_N)}$ impinges on the interface $z = x_N$. We wish to find the reflected wave in $z > x_N$ and the transmitted wave in $z < x_0$.

In theory the solution of this problem is reasonably straightforward. In practice, it can be extremely messy as far as arithmetic and algebra are concerned, especially when N is large. It was essentially this unpleasant feature that led Stokes to devise an imbedding type approach. He assumed the answers to be known for the case given, call them ρ_N and τ_N, then augmented the system by adding still another medium in $x_N < z < x_{N+1}$, and derived relationships among ρ_N, τ_N, ρ_{N+1}, and τ_{N+1}. Since the answer to the problem is known in the two medium case, $N = 0$, the resulting recursion relations may be used to obtain the reflection and transmission functions for any N.

Stokes' reasoning was physical, and we shall follow it reasonably closely. Basic to the analysis is the fact that one may trace a single wave, accounting for its reflection and transmission at each interface just ahead of it as if the material on each side of that interface is infinite and homogeneous and characterized by the wave numbers pertaining to the media just ahead of and just behind the wave. This observation, which seems highly physical, is really a statement of the fact that waves cannot interact with one another. This, in turn, is a result of the linearity of the wave equation. The observation is sometimes referred to as the "principle of localization."

In the next section we shall give the detailed "physical" argument. The reader will note that it has a great deal in common with the arguments used in Chapter 1. Later, we shall obtain the same type of result using purely analytical methods.

5. RESOLUTION OF THE MULTIMEDIUM PROBLEM BY "WAVELET COUNTING"

We consider the augmented medium as presented in the previous section. The impinging wave is now

$$e^{-ik_{N+2}(z-x_{N+1})}, \tag{6.6}$$

so that the solution for $z < x_0$ is

$$\psi = \tau_{N+1} e^{-ik_0(z-x_0)}, \qquad \tau_{N+1} \equiv \tau(x_{N+1}), \tag{6.7}$$

while that for $z > x_{N+1}$ can be written

$$\psi = e^{-ik_{N+2}(z-x_{N+1})} + \rho_{N+1} e^{ik_{N+2}(z-x_{N+1})}, \ \rho_{N+1} \equiv \rho(x_{N+1}). \tag{6.8}$$

We shall concentrate our attention on ρ_{N+1} for the present.

The impinging wave, upon arrival at $z=x_{N+1}$ is partly reflected and partly transmitted. The reflected portion makes an immediate contribution to ρ_{N+1}. According to Section 3 and the "principle of localization," this contribution may be computed from Eq. (6.5) and amounts to

$$\frac{k_{N+2}-k_{N+1}}{k_{N+2}+k_{N+1}}. \tag{6.9}$$

Similarly, the portion transmitted is

$$\frac{2k_{N+2}}{k_{N+2}+k_{N+1}}. \tag{6.10}$$

The wave impinging on the interface at $z=x_N$ is thus

$$\frac{2k_{N+2}}{k_{N+2}+k_{N+1}}e^{-ik_{N+1}(x_N-x_{N+1})}. \tag{6.11}$$

Equation (6.11) gives the strength of the "source" seen by the medium extending from $z=-\infty$ to $z=x_N$. This medium responds by reflecting a wave

$$\frac{2k_{N+2}}{k_{N+2}+k_{N+1}}e^{-ik_{N+1}(x_N-x_{N+1})}\rho_N, \tag{6.12}$$

the wave emergent at the interface $z=x_N$. The wave which reaches $z=x_{N+1}$ is

$$\frac{2k_{N+2}}{k_{N+2}+k_{N+1}}e^{-ik_{N+1}(x_N-x_{N+1})}e^{+ik_{N+1}(x_{N+1}-x_N)}\rho_N. \tag{6.13}$$

Again, this is in part reflected and in part transmitted into the medium to the right of $z=x_{N+1}$. This latter phenomenon makes a contribution to the overall wave reflection from the medium of

$$\frac{2k_{N+2}}{k_{N+2}+k_{N+1}}e^{2ik_{N+1}\Delta_{N+1}}\rho_N\frac{2k_{N+1}}{k_{N+2}+k_{N+1}}; \; \Delta_{N+1}=(x_{N+1}-x_N), \tag{6.14}$$

as may be seen by again employing Eq. (6.5).

The reflected part of the wave must now again be traced back to $z=x_N$ where reflection once more occurs, producing a right-moving wave, which upon reaching $z = x_{N+1}$ again contributes to the reflection coefficient:

$$\frac{2k_{N+2}}{k_{N+2}+k_{N+1}} \cdot \frac{2k_{N+1}}{k_{N+2}+k_{N+1}} e^{4ik_{N+1}\Delta_{N+1}} \left(\frac{k_{N+1}-k_{N+2}}{k_{N+2}+k_{N+1}} \right) \rho_N^2. \quad (6.15)$$

Obviously this counting procedure must be continued. The contribution involving $(2j+1)$ multiple reflections is

$$\frac{2k_{N+2}}{k_{N+2}+k_{N+1}} \cdot \frac{2k_{N+1}}{k_{N+2}+k_{N+1}} e^{2i(j+1)k_{N+1}\Delta_{N+1}} (-1)^j \left(\frac{k_{N+2}-k_{N+1}}{k_{N+2}+k_{N+1}} \right)^j \rho_N^{j+1}. \quad (6.16)$$

Thus the total reflection at $z = x_{N+1}$ is

$$\rho_{N+1} = \frac{k_{N+2}-k_{N+1}}{k_{N+2}+k_{N+1}} + \frac{4k_{N+2}k_{N+1}}{(k_{N+2}+k_{N+1})^2} \sum_{j=0}^{\infty} e^{2i(j+1)k_{N+1}\Delta_{N+1}} (-1)^j \left(\frac{k_{N+2}-k_{N+1}}{k_{N+2}+k_{N+1}} \right)^j \rho_N^{j+1}. \quad (6.17)$$

It is easy to see that the series is convergent and the sum may be written rather simply as

$$\rho_{N+1} = \frac{k_{N+2}-k_{N+1}}{k_{N+2}+k_{N+1}} + \frac{4k_{N+2}k_{N+1}}{(k_{N+2}+k_{N+1})^2} \frac{e^{2ik_{N+1}\Delta_{N+1}}\rho_N}{1+\left(\frac{k_{N+2}-k_{N+1}}{k_{N+2}+k_{N+1}} \right) e^{2ik_{N+1}\Delta_{N+1}}\rho_N}. \quad (6.18)$$

This result is equivalent to that obtained by Stokes [1].

The reader has doubtless noticed that we have accounted for all multiple internal reflections. This is in contrast to the "counting" procedures used in earlier chapters where only a relatively few such reflections had to be considered. The reason for the added complication here is that we have in no sense assumed any of the physical quantities "small." In the next section we shall use (6.18) to obtain results when such additional assumptions are made. At that time we shall see how the counting process above simplifies under such additional hypotheses.

Without entering into further detail we state the corresponding result for the transmission function, leaving the derivation to the reader as a problem:

$$\tau_{N+1}=\frac{2k_{N+2}}{k_{N+2}+k_{N+1}}\,\frac{e^{ik_{N+1}\Delta_{N+1}}\tau_N}{1+e^{2ik_{N+1}\Delta_{N+1}}\left(\dfrac{k_{N+2}-k_{N+1}}{k_{N+2}+k_{N+1}}\right)\rho_N}. \tag{6.19}$$

Since both ρ_0 and τ_0 are known, Eqs. (6.18) and (6.19) now allow one to compute the reflection and transmission functions for any number of media.

6. A CONTINUOUS MEDIUM PROBLEM

There is a strong temptation to attempt to generalize the results just obtained to cover the case of a medium in which the wave number varies in a continuous fashion. An obvious approach is to try to let the medium we have been studying remain of constant thickness while the total number of strata involved becomes infinite, with the maximum of $(x_{j+1}-x_j)$ going to zero. We shall carry out such a program, recognizing that there is a considerable amount of formalism involved. Later the same results will be obtained using a rigorous argument.

It will become apparent in our derivation that something more than simple continuity of the wave number will be needed for our result. We shall need over and over again to use the expansion

$$k(z+h)=k(z)+hk'(z)+o(h) \tag{6.20}$$

for $x_0<z<\infty$. In particular, this representation must hold at the right edge of the medium. This implies that we must imbed the material under study in a medium on the right which joins in a very smooth fashion—a joining that may be impossible to achieve in many cases. This requirement may seem to seriously limit the applicability of our results. We shall see eventually that it does not.

Let us examine the problem as it now stands in order to understand better what is taking place. We have the wave equation

$$\psi''(z)+k^2(z)\psi(z)=0, \qquad x_0<z<x, \tag{6.21}$$

where $k(z)$ satisfies (6.20) in $x_0<z<x$. We adjoin to this the equation

$$\psi''(z)+k_0^2\psi(z)=0, \qquad z<x_0, \tag{6.22}$$

where no restriction is placed on k_0 save that it be positive. We also adjoin

$$\psi''(z)+k_r^2\psi(z)=0, \qquad z>x, \tag{6.23}$$

where k_r is constant, $k_r=k(x-0)$, and (6.20) holds for $x_0<z<\infty$; in particular it holds at $z=x$. Assuming this is possible we now introduce a left-moving wave at $z=\infty$. This wave impinges on the medium at $z=x$, creating a reflected wave in the region $z>x$, and a wave in $x_0<z<x$. This latter wave produces a wave disturbance at $z=x_0$, and thus a left-moving wave is created in the region with wave number k_0. Hence

$$\psi(z)=e^{-ik_r(z-x)}+\rho(x)e^{ik_r(z-x)}, \qquad z>x;$$

$$\psi(z)=\tau(x)e^{-ik_0(z-x_0)}, \qquad z<x_0. \tag{6.24}$$

We now ask for the reflection coefficient $\rho(x)$ and the transmission coefficient $\tau(x)$.

The problem is evidently well posed provided the imbedding can be achieved, a very serious restriction, as we mentioned earlier, on the function $k(z)$. To avoid the major part of the difficulty we consider a somewhat bolder imbedding. In the region $z>x$ we introduce a medium of quite arbitrary wave number except that we ask that it, too, satisfy (6.20). Moreover, we require that the new medium join on to the given one in such a way that (6.20) is satisfied at $z=x$. Now if we introduce a wave in the usual way at $z=\infty$ it is likely to be completely distorted by the time it reaches $z=x$. However, a rereading of the analysis of the previous section reveals that only the form of this left-moving wave immediately to the right of $z=x$ is of any consequence. We can so arrange matters that *locally* this incoming wave is like $e^{-ik(x)(z-x)}$, which is to say that a unit wave impinges at $z=x$. The resulting reflected wave may be considered as looking *locally* like $\rho(x)e^{+ik(x)(z-x)}$. Thus at $z=x$ we have the condition

$$\psi(x)=1+\rho(x). \tag{6.25}$$

Equation (6.25) is not enough. A condition of continuity of the derivative of ψ must ultimately be imposed. We have argued that immediately to the right of $z=x$ the wave may be thought of as

$$\psi(z)=e^{-ik(x)(z-x)}+\rho(x)e^{ik(x)(z-x)}. \tag{6.26}$$

Differentiation of Eq. (6.26) with respect to z is invalid, for it is only a local representation. If such differentiation were legitimate then the derivative condition at $z=x$ would be

$$\psi'(x) = -ik(x)(1-\rho(x)). \tag{6.27}$$

We now accept (6.27) as a condition that we shall impose on $\rho(x)$. In effect, this partly *defines* the new function ρ. The definition is completely consistent with all previous work, and we have really defined a much more general reflection function than the one we have been using in this chapter. We shall show in the following that no further conditions must be specified. The function ρ is now completely determined.

Finally, we note that the given medium itself may be considered in part as the material in which the imbedding takes place. Thus, if $x_0 < x' < x$ then all the arguments just used may be made at the point $z=x'$, provided only that (6.20) is satisfied. We henceforth take this as a standard assumption concerning $k(z)$.

Now that this understanding of the imbedding process has been achieved we may begin the actual formal calculations. We augment the material at $z=x$ by a small amount, extending it to $z=x+\Delta$. In Eq. (6.18) write

$$\begin{aligned} k_{N+1} &= k(x), \qquad k_{N+2} = k(x+\Delta) = k(x) + k'(x)\Delta + o(\Delta), \\ \rho_N &= \rho(x), \qquad \rho_{N+1} = \rho(x+\Delta). \end{aligned} \tag{6.28}$$

Thus, for example,

$$\frac{k_{N+2}-k_{N+1}}{k_{N+2}+k_{N+1}} = \frac{k'(x)}{2k(x)}\Delta + o(\Delta). \tag{6.29}$$

A routine but somewhat tedious analysis of (6.18) now gives

$$\rho(x+\Delta) = \frac{k'(x)}{2k(x)}\Delta + \rho(x)\left[1 + 2ik(x)\Delta - \frac{k'(x)}{2k(x)}\Delta\rho(x) + o(\Delta)\right], \tag{6.30}$$

and passage to the limit yields

$$\rho'(x) = \frac{k'(x)}{2k(x)} + 2ik(x)\rho(x) - \frac{k'(x)}{2k(x)}\rho^2(x). \tag{6.31}$$

Since a genuine discontinuity of wave numbers may well exist at $z=x_0$ the condition there is

$$\rho(x_0) = \frac{k(x_0+0)-k_0}{k(x_0+0)+k_0}. \tag{6.32}$$

This system may now be integrated to find the reflection coefficient for any meaningful x value.

A similar calculation gives

$$\tau'(x)=\left[\frac{k'(x)}{2k(x)}+ik(x)-\frac{k'(x)}{2k(x)}\rho(x)\right]\tau(x),$$

$$\tau(x_0)=\frac{2k(x_0+0)}{k(x_0+0)+k_0}. \tag{6.33}$$

It may now be noted, by referring to (6.17) and (6.29), that for these computations it is not necessary in the counting procedure of the previous section to keep track of all multiple reflections. Obviously, only the terms corresponding to $j=0$ and $j=1$ in the sum (6.17) are needed. This is in complete agreement with the kind of observations and calculations made in earlier chapters.

We have frequently called attention to the fact that the arguments of this section are quite formal. Stokes presented such a treatment, with full awareness of its limitations. In the next section we shall make all of these ideas much more precise. Indeed, the arguments of this section could have been omitted completely. However, we feel that the insight provided is vital and compensates for the lack of mathematical rigor inherent in the presentation.

7. AN ANALYTICAL APPROACH TO THE CONTINUOUS MEDIUM PROBLEM

It is not too difficult to put the reasoning of the preceding section on a completely analytical basis. We shall now do this, thus placing our results on a much firmer foundation. The reader will note, however, that the "physical feeling" for the problem is somewhat obscured by the more elegant mathematics.

We consider the system

$$\psi''(z)+k^2(z)\psi(z)=0, \qquad x_0\leqslant z<\infty;$$

$$\psi(x_0)=\tau(x), \qquad \psi'(x_0)=-ik_0\tau(x); \tag{6.34}$$

$$\psi(x)=1+\rho(x), \qquad \psi'(x)=-ik(x)[1-\rho(x)], \qquad x_0<x,$$

where $k'(z)$ is at least piecewise continuous on $x_0<z<\infty$. Equation (6.34) combines all of the various conditions imposed in the last section. The restriction on $k(z)$ is considerably weaker than that previously stipulated.

Let $u_1(z)$ and $u_2(z)$ be those two linearly independent solutions of the wave equation satisfying

$$u_1(x_0)=1, \qquad u_1'(x_0)=0,$$

$$u_2(x_0)=0, \qquad u_2'(x_0)=1. \tag{6.35}$$

Writing

$$\psi(z)=c_1u_1(z)+c_2u_2(z), \qquad x_0\leqslant z, \tag{6.36}$$

we use (6.34) to compute

$$c_1=\tau(x), \qquad c_2=-ik_0\tau(x), \tag{6.37}$$

leading to

$$1+\rho(x)=\tau(x)[u_1(x)-ik_0u_2(x)],$$

$$-ik(x)[1-\rho(x)]=\tau(x)[u_1'(x)-ik_0u_2'(x)]. \tag{6.38}$$

Equation (6.38) readily yields

$$\rho(x)=\frac{\phi'(x)+ik(x)\phi(x)}{-\phi'(x)+ik(x)\phi(x)}, \tag{6.39}$$

where

$$\phi(x)=u_1(x)-ik_0u_2(x). \tag{6.40}$$

The form of (6.39) and the fact that ϕ satisfies a second order linear differential equation indicates that the function $\rho(x)$ must satisfy a Riccati equation. Elementary but tedious calculation gives

$$\rho'(x)=2ik(x)\rho(x)+\frac{k'(x)}{2k(x)}[1-\rho^2(x)], \tag{6.41}$$

in complete agreement with Eq. (6.31). If one solves for τ, he readily obtains Eq. (6.33).

All of the operations carried out in the previous paragraph are rigorous provided that the denominator in (6.39) does not vanish. We leave this matter to the problems. Thus we have a completely analytic derivation of those results obtained through physical reasoning in the previous section.

8. THE W. K. B. METHOD

Let us digress briefly to discuss a classical approximate solution to the wave equation, usually referred to as the W. K. B. approximation. The result actually goes back to Liouville (for an extensive discussion of this important method, together with some of its history, we refer the reader to [2]). If in the differential equation of (6.34) we write

$$s=\int_{x_0}^{z} k(t)\,dt,$$

we obtain

$$\frac{d^2}{ds^2}\hat{\psi}(s)+\frac{k'(z)}{k^2(z)}\frac{d\hat{\psi}}{ds}+\hat{\psi}(s)=0, \qquad \hat{\psi}(s)\equiv\psi(z). \tag{6.42}$$

The middle term may be eliminated in the usual fashion by use of the transformation,

$$\phi(s)=\sqrt{k(z)}\,\hat{\psi}(s), \tag{6.43}$$

yielding finally

$$\frac{d^2\phi(s)}{ds^2}+\phi(s)\left[1-\frac{1}{2\sqrt{k(z)}}\frac{d}{dz}\left(\frac{k'(z)}{k^{5/2}(z)}\right)-\frac{1}{2}\left(\frac{k'(z)}{k^2(z)}\right)^2\right]=0. \tag{6.44}$$

If now $k'(z)$ is in some sense small it is reasonable to expect that the solution to (6.44) is close to the solution of the simpler equation

$$\frac{d^2w(s)}{ds^2}+w(s)=0. \tag{6.45}$$

Thus we anticipate the approximation

$$\psi(z)\approx\frac{e^{\pm is}}{\sqrt{k(z)}}=\frac{1}{\sqrt{k(z)}}\exp\left(\pm i\int_{x_0}^{z}k(t)\,dt\right). \tag{6.46}$$

This analysis may be made rigorous under various assumptions concerning k'/k.

Let us turn to Eq. (6.33) and approximate it by

$$\tau_0'(x)=\left[\frac{k'(x)}{2k(x)}+ik(x)\right]\tau_0(x). \tag{6.47}$$

That is, we assume that ρ is identically zero. There is no internal reflection. Physically this situation will be approximated when $k(z)$ is a slowly varying function. The more rapidly k changes locally, the more the medium acts as if there is a local discontinuity in the wave number, and such a discontinuity is the equivalent of a true interface.

Equation (6.47) can be solved easily:

$$\tau_0(x)=\tau_0(x_0)\sqrt{\frac{k(x)}{k(x_0)}}\ \exp\left(i\int_{x_0}^{x}k(t)\,dt\right). \tag{6.48}$$

If we take k to be continuous at x_0, then (6.33) shows that $\tau(x_0)=1$. We require $\tau_0(x_0)=1$. Finally, from (6.34) we find

$$\psi(x_0)\approx\sqrt{\frac{k(x)}{k(x_0)}}\ \exp\left(-i\int_{x}^{x_0}k(t)\,dt\right), \tag{6.49}$$

where we have written the approximation sign to provide a reminder that Eq. (6.47) is not exact. We observe that the result is equivalent to the W. K. B. approximation in this case. Thus we can proceed as if the classical W. K. B. approach provides that wave which would exist if there were no internal reflections.

9. THE BREMMER SERIES

In the late 1930's H. Bremmer noticed that the W. K. B. approximation may be considered to represent the first term in an infinite series that can be constructed by keeping account of all multiple internal reflections [3,4]. Later investigations revealed that the formal series which he obtained really converges to a solution of the wave equation in a wide variety of cases [5,6]. Since the last several sections have been based on analyses that have taken account of such multiple reflections, it is reasonable to expect that we should be able to derive Bremmer's results from the equations we have already obtained and without further detailed "counting." This program we now carry out.

Let us focus attention on the approximate W. K. B. wave. At any position t, $x_0\leqslant t\leqslant x$, this transmitted wave may be reflected, producing a right-moving wave. Unfortunately, the equations we have ordinarily used for the reflection coefficient in no way involve the transmission function. To remedy this we multiply (6.33) by ρ, solve for ρ^2 and use the resulting expression in (6.31). The equation obtained is

$$\rho'(x)+\rho(x)\left[\frac{k'(x)}{2k(x)}-ik(x)-\frac{\tau'(x)}{\tau(x)}\right]=\frac{k'(x)}{2k(x)}. \tag{6.50}$$

We now use for τ in this equation the function obtained by W. K. B. (see (6.48)). Writing this function as τ_0 and the new reflection function as ρ_1, we have

$$\rho_1'(x)+\rho_1(x)\left[\frac{k'(x)}{2k(x)}-ik(x)-\frac{\tau_0'(x)}{\tau_0(x)}\right]=\frac{k'(x)}{2k(x)} \tag{6.51}$$

with solution at any position z, $x_0 \leqslant z \leqslant x$,

$$\rho_1(z)=-\frac{\tau_0(z)}{\sqrt{k(z)}}\int_z^{x_0}\frac{k'(t)}{2\sqrt{k(t)}}(\tau_0(t))^{-1}\exp\left(-i\int_z^t k(s)\,ds\right)dt. \tag{6.52}$$

Here we have again supposed that $k(z)$ is continuous at x_0 so that $\rho_1(x_0)=0$.

Equation (6.52) in its present form is difficult to understand. Let us recall that $\tau_0(t)$ would be the value of the W. K. B. wave at x_0 *if* the input at t were a unit wave. However, the input at t is actually the W. K. B. wave there. Denoting this wave by ψ_0 we see that

$$\tau_0(t)\psi_0(t)=\psi_0(x_0)=\tau_0(x). \tag{6.53}$$

Thus (6.52) may be rewritten

$$\rho_1(z)=-\frac{\tau_0(z)}{\tau_0(x)\sqrt{k(z)}}\int_z^{x_0}\frac{k'(t)}{2\sqrt{k(t)}}\psi_0(t)\exp\left(-i\int_z^t k(s)\,ds\right)dt. \tag{6.54}$$

Again $\rho_1(z)$ would be the right moving wave, ψ_1, at z if the input were unity. Since the input is actually $\psi_0(z)$ we have

$$\psi_0(z)\rho_1(z)=\psi_1(z). \tag{6.55}$$

Finally, because (6.53) holds at $t=z$,

$$\tau_0(z)\psi_0(z)=\psi_0(x_0)=\tau_0(x), \tag{6.56}$$

and Eq. (6.54) assumes the much more suggestive form

$$\psi_1(z)=-\frac{1}{2\sqrt{k(z)}}\int_z^{x_0}\frac{k'(t)}{\sqrt{k(t)}}\psi_0(t)\exp\left(-i\int_z^t k(s)\,ds\right)dt. \tag{6.57}$$

The wave ψ_1 impinges on the medium extending from z to x, and a reflected wave is to be expected at z. Unfortunately, the wave ψ_1 is traveling to the right and is an input at the left side of this medium. All our equations have been derived for exactly the opposite situation. The derivation of the appropriate formulas is straightforward but tedious, and thus we leave the details as a problem. The result analogous to (6.50) is

$$\hat{\rho}'(z)+\hat{\rho}(z)\left[\frac{k'(z)}{2k(z)}+ik(z)-\frac{\hat{\tau}'(z)}{\hat{\tau}(z)}\right]=\frac{k'(z)}{2k(z)}, \tag{6.58}$$

where we have written $\hat{\rho}$ and $\hat{\tau}$ to indicate that these are the reflection and transmission functions pertaining to the wave input on the left. Following the treatment we have just presented we find

$$\psi_2(z)=\frac{1}{2\sqrt{k(z)}}\int_x^z\frac{k'(t)}{\sqrt{k(t)}}\psi_1(t)\exp\left(i\int_z^t k(s)\,ds\right)dt. \tag{6.59}$$

Here ψ_2 is a left-moving wave impinging at z.

The iterative scheme is now clear and we obtain a sequence of right- and left-moving waves

$$\psi_{2n+1}(z)=-\frac{1}{2\sqrt{k(z)}}\int_z^{x_0}\frac{k'(t)}{\sqrt{k(t)}}\psi_{2n}(t)\exp\left(-i\int_z^t k(s)\,ds\right)dt, \tag{6.60a}$$

$$\psi_{2n}(z)=\frac{1}{2\sqrt{k(z)}}\int_x^z\frac{k'(t)}{\sqrt{k(t)}}\psi_{2n-1}(t)\exp\left(i\int_z^t k(s)\,ds\right)dt. \tag{6.60b}$$

The physics of the situation suggests strongly that the solution ψ of the problem posed in Eq. (6.34) should be simply

$$\psi(z)=\sum_{n=0}^{\infty}\psi_n(z). \tag{6.61}$$

This is the Bremmer series. As mentioned earlier it may be verified that this expansion is indeed valid in a large number of cases. An investigation in depth is not appropriate here and we refer the reader to the original works and to the problems.

10. ANOTHER IMBEDDING

As we have taken great pains to point out, the imbedding that has been used thus far in this chapter is of a different nature from that employed in the early portions of this book. We now ask if the early results may in any way be used in the study of wave propagation.

Consider the material as extending again from x_0 to x, but let the wave number in the region $z<x_0$ be k_0 and that in $z>x$ be k_1 with no restrictions concerning the "joining" of the media at $z=x$. Moreover, we shall ask that $k(z)$, $x_0<z<x$, be only piecewise continuous, and, of course, positive. Thus the smoothness of the wave number that has been very important to our analysis is no longer required.

Let a left-moving wave impinge on the medium from $z=+\infty$. As in the previous arguments the wave in the region $z>x$ is of the form

$$\psi(z)=e^{-ik_1(z-x)}+\rho(x)e^{+ik_1(z-x)}, \tag{6.62}$$

while that in $z<x_0$ is

$$\psi(z)=\tau(x)e^{-ik_0(z-x_0)}. \tag{6.63}$$

In Eqs. (6.62) and (6.63) we continue to write the reflection and transmission functions as ρ and τ, since their physical meaning is as before. We expect, however, that they will satisfy different equations than those found in earlier sections of this chapter. Indeed, since we have made no assumptions about the differentiability of $k(z)$, equations like (6.31) are not to be anticipated.

Our next effort is to obtain a formulation analogous to that involved in simple transport models. To do this we try to introduce functions u and v in such a way that u suggests a wave moving to the right and v suggests one moving to the left. Define

$$u(z)=\frac{1}{2ik_0}(\psi'(z)+ik_0\psi(z)),$$

$$v(z)=-\frac{1}{2ik_1}(\psi'(z)-ik_1\psi(z)). \tag{6.64}$$

Notice that in the medium with wave number k_0 ($z<x_0$) a right-moving wave has the form e^{ik_0z}. Thus in that region $u(z)\equiv e^{ik_0z}$. An analogous remark holds for $v(z)$ in the medium of wave number k_1. Thus this choice of u and v is physically sensible.

Some detailed calculations with $\psi''(z)+k^2(z)\psi(z)=0$ reveal that

$$u'(z)=\left[\frac{1}{ik_0(k_0+k_1)}\right]\{-k_0[k^2(z)+k_0k_1]u(z)-k_1[k^2(z)-k_0^2]v(z)\}, \tag{6.65a}$$

$$-v'(z)=\left[\frac{1}{ik_1(k_0+k_1)}\right]\{-k_0[k^2(z)-k_1^2]u(z)-k_1[k^2(z)+k_0k_1]v(z)\}. \tag{6.65b}$$

Since at the interfaces we have (recall (6.34))

$$\begin{aligned}\psi(x_0)&=\tau(x), & \psi'(x_0)&=-ik_0\tau(x),\\ \psi(x)&=1+\rho(x), & \psi'(x)&=-ik_1(1-\rho(x)),\end{aligned} \tag{6.66}$$

the functions u and v satisfy

$$u(x_0)=0, \qquad v(x)=1. \tag{6.67}$$

We now obviously have a system of precisely the form of (2.16). The equation for $r(x)$ may be written down automatically using (2.21a):

$$\begin{aligned}r'(x)=\frac{1}{i(k_0+k_1)}\bigg\{\frac{k_1}{k_0}[k_0^2-k^2(x)]-2[k^2(x)+k_0k_1]r(x)\\ +\frac{k_0}{k_1}[k_1^2-k^2(x)]r^2(x)\bigg\}, \quad r(x_0)=0.\end{aligned} \tag{6.68}$$

The reader should be careful to note that this is indeed an equation for the function r, where $r(x)=u(x)$, and *not* an equation for $\rho(x)$. However, in view of the relationships available from (6.64), (6.66), and (6.67), it is relatively easy to see that

$$\rho(x)=\frac{2k_0}{k_1+k_0}r(x)+\frac{k_1-k_0}{k_1+k_0}, \tag{6.69}$$

and so

$$\begin{aligned}\rho'(x)=\frac{(-1)}{2ik_1}\{[k^2(x)-k_1^2]+2[k^2(x)+k_1^2]\rho(x)\\ +[k^2(x)-k_1^2]\rho^2(x)\}, \qquad \rho(x_0)=\frac{k_1-k_0}{k_1+k_0}.\end{aligned} \tag{6.70}$$

A similar analysis produces an equation for τ:

$$\tau'(x) = \frac{(-1)}{2ik_1}\{[k^2(x)+k_1^2]+[k^2(x)-k_1^2]\rho(x)\}\tau(x),$$

$$\tau(x_0) = \frac{2k_1}{k_1+k_0}. \tag{6.71}$$

It is interesting to observe that whenever the joining at the interface $z = x$ can be accomplished in a smooth manner, in the sense of the previous sections, the solutions to Eq. (6.70) and (6.71) must agree with the solutions to (6.31) and (6.33). This is by no means obvious from the form of the equations themselves.

11. SUMMARY

In this chapter we have applied the method of invariant imbedding to the problem of wave propagation. Physical requirements have led to a new and extended type of imbedding. Results obtained from this new imbedding have yielded quite naturally both the famous W. K. B. method and the extension of that method due to Bremmer. In addition, we have seen that the devices originally developed in Chapters 1 and 2 may also be used, although the equations obtained for the reflection and transmission functions depend very strongly on the imbedding utilized.

Again numerous questions arise. Can the method be employed to study a plane wave obliquely impinging on a "stratified" medium? The answer is in the affirmative, although we leave the details to the problems. The matter of polarized waves also demands investigation. This leads to a vector wave equation, which may be dealt with using the methods of the early chapters. Details of the corresponding analysis when the new type of imbedding is used have not been carried through in detail. The matter of spherical waves and the like we shall also leave untouched here. Indeed, *all* of our results thus far have involved geometries which may be described as "plane parallel." Although some efforts have been made to study the method of invariant imbedding in other geometries, the difficulties are challenging.

PROBLEMS

1. Consider the "three medium" wave problem in which the medium extending from $-\infty$ to x_0 has wave number k_0, that between x_0 and x_1 has wave number k_1, and that from x_1 to $+\infty$ has wave number k_2. Let a left-moving wave start at $+\infty$. Solve the resulting problem completely from first principles, paying particular

attention to reflection and transmission at $z=x_0$ and $z=x_1$. Check your results with the general ones of Section 5.

2. Discuss the "principle of localization" both from a physical viewpoint and as a consequence of the structure of the wave equation.

3. Prove that the series in (6.17) converges.

4. Derive (6.19) using the general ideas employed in obtaining (6.18). Obtain (6.33) by at least two methods.

5. Derive Eqs. (6.68), (6.70), and (6.71).

6.* Study the relationship between the solutions to (6.70) and (6.71) and those of (6.31) and (6.33) in the case in which the joining can be accomplished at the interface $z=x$ in a "smooth" manner.

7. Consider the case of a plane wave impinging obliquely on a slab of material of the sort studied in this chapter. Show that the various equations and functions that we have investigated can be generalized rather easily to include this case.

8. Obtain Eqs. (6.60) in detail. Suggest and verify some sufficient conditions for the convergence of the Bremmer series.

9. Derive (6.38).

10. Discuss the possibility of a phenomenon analogous to criticality in the case of wave propagation.

11. Perform the "elementary but tedious" calculations necessary to get (6.41). Use the same method to obtain (6.33).

12. Prove that the denominator in (6.39) cannot vanish.

13. Derive the ρ and τ equations for input at the left of the medium. These are used in the derivation of the Bremmer series, where the functions are denoted by $\hat{\rho}$ and $\hat{\tau}$.

14.* Develop an invariant imbedding theory for plane polarized waves.

15. Consider a wave traveling through a medium with wave number $k(z)$. Suppose that in passing through a stratum $(z,z+\Delta)$ there is a probability $p(z)\Delta+o(\Delta)$ that the wave will be absorbed but in return will produce two waves of the same amplitude and phase, one traveling to the left and one to the right. The probability that such a "fission" will not occur in the stratum is, of course, $1-p(z)\Delta+o(\Delta)$. Show that the reflection function for a right-moving wave impinging on the stratum (z,b) is

$$\frac{d\hat{\rho}}{dz}=\frac{k'(z)}{2k(z)}-p(z)-2ik(z)\hat{\rho}(z)-\left(\frac{k'(z)}{2k(z)}+p(z)\right)\hat{\rho}(z)^2.$$

What is the initial condition on $\hat{\rho}$?

16.* The Bremmer series can actually be obtained as the Neumann series of a suitable linear integral equation. Find that equation.

17.* The problem of convergence of the Bremmer series is a reasonably difficult one. Some criteria have been obtained in Problem 8. Now let $x_0=-\infty$. Using the approach suggested in Problem 16 try to show that the series will converge if

$\int_{-\infty}^{x}|k'(t)/k(t)|\,dt \leqslant \pi$. Prove that π cannot be replaced by any larger number. (See F. V. Atkinson, "Wave Propagation and the Bremmer Series," *J. Math. Anal. Appl.* **1**, 1960, 255–276.)

REFERENCES

1. G. C. Stokes, *On the Intensity of the Light Reflected or Transmitted Through a Pile of Plates*, Mathematical and Physical Papers, **2**, Cambridge Univ. Press, London, 1883.
2. R. Bellman, *Perturbation Techniques in Mathematics, Physics, and Engineering*, Holt, Rinehart, and Winston, New York, 1964.
3. H. Bremmer, in *Handelingen Natuur en Geneeskundig Congress*, Nijmiegen, p. 88, Ruygrok, Haarlem, 1939.
4. H. Bremmer, "The WKB Approximation as the First Term of a Geometric-Optical Series," in *Theory of Electromagnetic Waves*, pp. 125–138, Wiley-Interscience, New York, 1951.
5. R. Bellman and R. Kalaba, "Functional Equations, Wave Propagation, and Invariant Imbedding," *J. Math. Mech.* **8**, 1959, 683–704.
6. F. V. Atkinson, "Wave Propagation and the Bremmer Series," *J. Math. Anal. Appl.* **1**, 1960, 255–276.

7

TIME-DEPENDENT PROBLEMS

1. INTRODUCTION

Up to the present point in this book we have carefully avoided problems involving time-dependence. Only in certain of the models studied in the chapter on random walk was time included, and there it played a relatively minor role. Indeed, in the very simple transport problems investigated in the initial chapters, considerable care was taken to point out that time-independence was assumed, and only the so-called stationary state case was under study. Yet in many areas of mathematical physics phenomena that are dependent upon time are of great interest. It is now appropriate to confront this problem.

There has been good reason for our avoidance of time considerations: the mathematics involved often becomes quite complicated. This is, of course, a well-known fact in classical applied mathematics. In the invariant imbedding framework the situation originally seemed even worse. During the early days of the development of the theory, when interest was concentrated on simple transport models and almost all results were obtained by particle-counting, investigation of even the easiest time-dependent problems was so complicated that the resulting equations were highly suspect. When the counting approach gave way to more analytical methods in problems not involving time it was only natural to seek a device that could reduce time-dependent problems to equivalent stationary state problems. Such a device (actually one of many) has long been known in classical mathematics–the Laplace transform. As we shall see, transform methods are as powerful in handling invariant imbedding problems as they are in resolving questions posed more classically.

The simplification can be somewhat illusory, as is almost always the case when the transform method is employed. For, in the last analysis, it is always necessary to invert back to the original time domain. Although in many instances this can be done, the structure of the resulting equations is often such as to make them quite intractable analytically. In some cases, however, it has proved feasible and valuable to carry out the necessary inversion of the imbedding equations by numerical means.

In this chapter we shall discuss enough time-dependent problems to give the reader a general idea of the attacks that have proved successful. Partly as a matter of historical interest we shall begin with a particle-counting procedure applied to a one-dimensional transport problem. The complexities involved are such that the reader will likely be convinced that an easier way is definitely desirable. The counting procedure will then be verified and generalized by the transform technique to which we have referred. It will become clear how a wide class of transport models in which time enters in an explicit way can be handled by the imbedding method provided the corresponding time-independent problem is amenable to the method.

Having achieved some understanding of how to treat time-dependence in the transport case, we shall branch out and consider other problems of mathematical physics in which time plays a major role. Thus, such areas as wave propagation and diffusion theory are readily investigated, always provided, of course, that the underlying geometry is sufficiently simple.

2. A TIME-DEPENDENT TRANSPORT PROBLEM–PARTICLE-COUNTING APPROACH

Let us first derive the time-dependent analog of the simple transport equation (1.8) and (1.9) of Chapter 1. We may now allow the quantities σ, f, and b to depend on time as well as on position. Thus $\sigma=\sigma(z,t)$, and so on, and obviously the u and v functions must also be time-dependent. We define

$$u(z,t) = \text{the expected density of particles moving to the right at position } z \text{ at time } t. \tag{7.1}$$

The function $v(z,t)$ is defined similarly for left-moving particles.

A bit of care must be used in interpreting (7.1). We mean that u is that function such that at time t the expected number of right-moving particles in the interval $z_1 \leqslant z \leqslant z_2$ is given by

$$\int_{z_1}^{z_2} u(z,t)\,dz, \tag{7.2}$$

while the expected number of particles passing the point z to the right in the time interval $t_1 \leqslant t \leqslant t_2$ is

$$c\int_{t_1}^{t_2} u(z,t)\,dt, \tag{7.3}$$

where c is the particle speed. [Note that the function $u(z)$ defined by (1.2) is essentially a flux and so contains a factor c automatically. This slight difference in definitions should not create any difficulty.]

It is interesting to give a somewhat different type of derivation for the corresponding transport equation from that used in earlier discussions. Let us suppose we are "riding along" a beam or stream of right-moving particles. The type of physical interactions described in Chapter 1 will result in both diminution and augmentation of this beam. A collision interaction removes a particle from the stream, but it also produces f particles augmenting the stream. At the same time, left-moving particles may experience collisions and produce right-moving particles that also "join" the beam on which we are riding. We ask: How does the density of this beam vary as a function of time? (The reader familiar with hydrodynamics will note that our approach is essentially that of Lagrange.)

If we denote by D/Dt the total time derivative we find, after a bit of thought, that

$$\frac{Du}{Dt}(z,t) = -c\sigma(z,t)u(z,t) + c\sigma(z,t)u(z,t)f(z,t) + c\sigma(z,t)v(z,t)b(z,t), \tag{7.4}$$

and also that

$$\begin{aligned}\frac{Du}{Dt}(z,t) &= \frac{\partial u(z,t)}{\partial t} + \frac{\partial u(z,t)}{\partial z}\cdot\frac{dz}{dt},\\ &= \frac{\partial u(z,t)}{\partial t} + \frac{\partial u(z,t)}{\partial z}c.\end{aligned} \tag{7.5}$$

Here we have taken into account the fact that z actually changes since we are "riding" with the beam. Thus

$$\frac{\partial u}{\partial z} + \frac{1}{c}\frac{\partial u}{\partial t} = \sigma(z,t)\{(f(z,t)-1)u(z,t) + b(z,t)v(z,t)\}. \tag{7.6a}$$

The corresponding v equation is

$$-\frac{\partial v}{\partial z} + \frac{1}{c}\frac{\partial v}{\partial t} = \sigma(z,t)\{b(z,t)u(z,t) + (f(z,t)-1)v(z,t)\}. \tag{7.6b}$$

Observe that in the time-independent case Eqs. (7.6) reduce to Eqs. (1.8) and (1.9).

These partial differential equations must be subjected to certain boundary and initial conditions. Let us suppose that at time $t=0$ there are no particles at all in the rod. Let a single particle be injected at the right end, $z=x$, at time $t=0$, and assume that no other particles enter the rod at any time. Thus,

$$cu(0,t)=0, \tag{7.6c}$$

$$cv(x,t)=\delta(t), \tag{7.6d}$$

$$u(z,0)=v(z,0)=0, \qquad 0<z<x. \tag{7.6e}$$

(The function $\delta(t)$ is the Dirac delta function, of which we shall say a bit more later in this chapter.)

To simplify our reasoning let us choose very special functions f, b, and σ; namely,

$$f(z,t)=1, \qquad b(z,t)=1, \qquad \sigma(z,t)=\sigma, \text{ a constant.} \tag{7.7}$$

Thus we are actually about to study the time-dependent version of the model which we first investigated in Chapter 1. Equations (7.6) now become

$$\frac{\partial u}{\partial z}+\frac{1}{c}\frac{\partial u}{\partial t}=\sigma v, \tag{7.8a}$$

$$-\frac{\partial v}{\partial z}+\frac{1}{c}\frac{\partial v}{\partial t}=\sigma u, \tag{7.8b}$$

together with the boundary and initial conditions stated in (7.6c)–(7.6e).

As a classical problem this is now well-posed. Having obtained this version, we wish to derive the corresponding invariant imbedding equation, still requiring that (7.7) hold. We begin our efforts by taking no direct cognizance of Eqs. (7.8) and use purely physical reasoning.

Let us define a reflection function as follows:

$cr(x,t,\tau)=$ the expected *total* number of particles emergent from the rod at the right end, $z=x$, up to time t due to the injection at $z=x$ of a single particle at time $\tau, 0\leqslant\tau<t$. (7.9)

As a matter of convenience we define $cr(x,t,\tau)=0$ for $t\leqslant\tau$. Note that

$c(\partial r/\partial t)$ is the *rate* of emergence of particles at time t. We seek an equation satisfied by the function $cr(x,t,\tau)$.

In the standard manner we increment the rod length by Δ. It turns out to be desirable to change both t and τ, and thus we concentrate on the function $cr[x+\Delta,t+(\Delta/c),\tau-(\Delta/c)]$.

Consider the particle entering the augmented rod at time $\tau-(\Delta/c)$. It may suffer a collision in passing through the interval $x\leqslant z\leqslant x+\Delta$. If this occurs a particle emerges at $z=x+\Delta$. This contribution is $\sigma\Delta+o(\Delta)$. At the same time a left-moving particle emerges from the collision and proceeds toward the *sub*-rod, $0\leqslant z\leqslant x$. In the event that the particle injected at time $\tau-(\Delta/c)$ has no collision in $x\leqslant z\leqslant x+\Delta$ it will reach $z=x$ at time τ. The net result of both of these possibilities is that at time τ the *sub*-rod "sees" an input of $1+o(\Delta)$ particles.

Now consider a time $t', 0\leqslant t'<t$. Particles emerge from $z=x$ at a *rate* $c(\partial r/\partial t')(x,t',\tau)$. Again these may or may not experience collisions in the interval $x\leqslant z\leqslant x+\Delta$. In either case the *rate* of emergence at $z=x+\Delta$ is $c(\partial r/\partial t')(x,t',\tau)+o(\Delta)$. However, in the event of a collision in $x\leqslant z\leqslant x+\Delta$ particles will be reinjected into the *sub*-rod, and these represent a new source seen by that *sub*-rod at time t'. The strength of this source is $\sigma\Delta c(\partial r/\partial t')(x,t',\tau)$ [to within a term of order $o(\Delta)$] and this results in a total of $\sigma\Delta c(\partial/\partial t')r(x,t',\tau)cr(x,t,t')$ particles emergent at $z=x$ by time t. Combining all these effects, and taking into account that any others contribute terms of order at most $o(\Delta)$, we obtain the relation

$$cr\left(x+\Delta,t+\frac{\Delta}{c},\tau-\frac{\Delta}{c}\right)=\sigma\Delta$$

$$+c\int_0^t\frac{\partial r}{\partial t'}(x,t',\tau)\,dt'+c^2\sigma\Delta\int_0^t\frac{\partial}{\partial t'}r(x,t',\tau)r(x,t,t')\,dt'+o(\Delta),\quad (7.10)$$

leading at once to the equation

$$\frac{\partial r}{\partial x}+\frac{1}{c}\frac{\partial r}{\partial t}-\frac{1}{c}\frac{\partial r}{\partial \tau}=\frac{\sigma}{c}+\sigma c\int_0^t\frac{\partial}{\partial t'}r(x,t',\tau)r(x,t,t')\,dt'.\quad (7.11)$$

It should be recalled that we have already required that

$$r(x,t,\tau)=0,\qquad t\leqslant\tau.\quad (7.11a)$$

In addition, we clearly have

$$r(0,t,\tau)=0.\quad (7.11b)$$

Equations (7.11) serve to define the physical problem completely. Nevertheless, the structure of the equations is so unusual that the existence of a solution is by no means obvious. This matter is considered in some detail for a much more complicated situation in [1, 2].

A considerable simplification of (7.11) is possible. Since all physical parameters in the problem are time-independent (indeed, they have been assumed constant) the function r must actually depend on the time *difference* $t-\tau$, and not on t and τ independently. Let us then define

$$c\tilde{r}(x,t-\tau)\equiv cr(x,t,\tau). \tag{7.12}$$

Then the system (7.11) may be rewritten*

$$\frac{\partial\tilde{r}}{\partial x}+\frac{2}{c}\frac{\partial\tilde{r}}{\partial T}=\frac{\sigma}{c}+\sigma c\int_0^T\frac{\partial\tilde{r}}{\partial t'}(x,t')\tilde{r}(x,T-t')\,dt', \tag{7.13}$$

$$\tilde{r}(x,T)=0,\qquad T\leqslant 0,\qquad x\geqslant 0, \tag{7.13a}$$

$$\tilde{r}(0,T)=0,\qquad T\geqslant 0, \tag{7.13b}$$

where $T=t-\tau$. This is a substantial simplification of equations (7.11). In fact, the convolution form of the integral term suggests at once the use of the Laplace transform. The problem can be completely analyzed and resolved in this way (see [1]).

The particle-counting method may be employed in considerably more complicated physical models provided the rod geometry is retained. Thus, σ, f and b may all be made fairly general functions of *both* z and t. In the event that σ, f, and b depend *only* upon z the function $\tilde{r}$ defined by Eq. (7.12) may still be used, leading to some improvement in the form of the resulting expressions. If any of the physical parameters σ, f, or b depends explicitly on time this simplification is no longer available. We leave the details of many of these matters to the reader (see Problems 1, 2, and 5).

The complexity of the particle-counting analysis even in the simple rod geometry suggests that efforts at extending this type of argument to more realistic cases may be quite difficult and that chances for error can be very great. For this reason we seek another approach to the time-dependent

*Up to this point we have made every effort in the text to denote scalar quantities and column vectors by lowercase letters, reserving uppercase letters for matrices. For purely typographical reasons it is now sometimes desirable to abandon that practice. The reader should be able to infer from the context when a quantity is a scalar and when it is a vector. In any ambiguous cases we shall make every effort to clarify the matter with appropriate remarks in the text.

transport problem already studied. Unfortunately, the scheme we develop is only applicable in case the physical parameters are themselves time-independent. However, since the concepts involved are simple and indicate how the imbedding analysis may be readily applied to many other types of problems, it is certainly worthwhile pursuing.

3. TIME-DEPENDENT TRANSPORT BY TRANSFORM TECHNIQUES

Let us return to the transport problem posed by Eqs. (7.8). It will be recalled that explicit use of these equations was not made in the previous section, since we turned at once to physical arguments. We shall now employ (7.8) directly, reducing the problem to a more recognizable form by means of the Laplace transform in the time variable.

Before proceeding, it is worthwhile to note that the quantity $c\tilde{r}(x,T)$ found in the previous section is actually the *total* number of particles expected to emerge from the rod by time T due to the injection of a single particle at time zero at the right end. The functions $cu(z,t)$ and $cv(z,t)$ give the expected numbers of particles passing the point z to the right and left (respectively) *each second* at time t. To obtain agreement in our results we therefore choose to define the functions

$$c\tilde{u}(z,T)=\int_0^T cu(z,t')\,dt', \qquad c\tilde{v}(z,T)=\int_0^T cv(z,t')\,dt'. \tag{7.14}$$

From the system (7.8) we readily obtain

$$\frac{1}{c}\frac{\partial \tilde{u}}{\partial T}(z,T)+\frac{\partial \tilde{u}}{\partial z}=\sigma\tilde{v}(z,T), \tag{7.15}$$

$$\frac{1}{c}\frac{\partial \tilde{v}}{\partial T}(z,T)-\frac{\partial \tilde{v}}{\partial z}=\sigma\tilde{u}(z,T),$$

$$c\tilde{u}(0,T)=0, \qquad T\geqslant 0, \tag{7.15a}$$

$$c\tilde{v}(x,T)=1, \qquad T>0, \tag{7.15b}$$

$$c\tilde{u}(z,0)=c\tilde{v}(z,0)=0, \qquad 0\leqslant z\leqslant x. \tag{7.15c}$$

It is now clear that

$$c\tilde{r}(x,T)\equiv c\tilde{u}(x,T), \tag{7.16}$$

and it is our hope to obtain equation (7.13) directly from (7.15).

Define

$$L_T\{c\tilde{u}(z,T)\} = \int_0^\infty e^{-sT} c\tilde{u}(z,T)\,dT = \tilde{U}(z,s),$$
$$L_T\{c\tilde{v}(z,T)\} = \tilde{V}(z,s). \tag{7.17}$$

Applying this Laplace transformation to (7.15) we easily find

$$\frac{d\tilde{U}}{dz}(z,s) = -\frac{s}{c}\tilde{U}(z,s) + \sigma\tilde{V}(z,s),$$

$$-\frac{d\tilde{V}}{dz}(z,s) = \sigma\tilde{U}(z,s) - \frac{s}{c}\tilde{V}(z,s), \tag{7.18}$$

$$\tilde{U}(0,s) = 0, \tag{7.18a}$$

$$\tilde{V}(x,s) = \frac{1}{s}. \tag{7.18b}$$

Equations (7.18), however, are in exactly the form studied in the early chapters of this book, except for the right side of (7.18b). Define

$$s\tilde{U}(z,s) = U(z,s), \qquad s\tilde{V}(z,s) = V(z,s). \tag{7.19}$$

Clearly, U and V still satisfy (7.18) and (7.18a) while (7.18b) is replaced by

$$V(x,s) = 1. \tag{7.18b$'$}$$

We may now apply the result in Eq. (3.25) to obtain

$$\frac{d}{dx}R(x,s) = \sigma - 2\frac{s}{c}R(x,s) + \sigma R^2(x,s),$$

$$R(0,s) = 0, \tag{7.20}$$

where $R(x,s) \equiv U(x,s)$. A further transformation,

$$\tilde{U}(x,s) = \tilde{R}(x,s) = \frac{1}{s}R(x,s), \tag{7.21}$$

when applied to (7.20) yields trivially

$$\frac{d\tilde{R}}{dx} = \frac{\sigma}{s} - \frac{2s}{c}\tilde{R}(x,s) + \sigma s\tilde{R}^2(x,s). \tag{7.22}$$

Finally, we note from (7.16) and (7.17) that

$$c\tilde{r}(x,T) = L_s^{-1}\{\tilde{R}(x,s)\}. \tag{7.23}$$

Elementary transform analysis now produces from (7.22) the desired result [compare (7.13)]:

$$\frac{\partial \tilde{r}}{\partial x} + \frac{2}{c}\frac{\partial \tilde{r}}{\partial T} = \frac{\sigma}{c} + \sigma c \int_0^T \frac{\partial \tilde{r}}{\partial t'}(x,t')\tilde{r}(x,T-t')\,dt'. \tag{7.24}$$

The side conditions found in the previous section are readily established here from the corresponding conditions placed on u and $\tilde{u}$.

4. A CRITIQUE OF THE FOREGOING

In the previous two sections we have derived invariant imbedding formulations of a very simple time-dependent transport model. The particle-counting scheme used in Section 2 is in some ways less elegant than the Laplace transform approach of Section 3. However, in the treatment of the earlier section it is possible to allow the physical parameters to depend on time (see Problems 1–3). An analysis of the preceding section quickly shows that the transform method fails completely in such a situation. [It should be noted, however, that variation of the parameters with z *only* would not make the transform method unusable (see Problem 5).] It must also be admitted that any imbedding formulation of other types of problems in which the coefficients in the classical equations are time-dependent is also likely to be quite difficult.

In Section 3, we employed a considerable amount of manipulation to obtain equations that produced knowledge about the total number of particles emergent from the system up to time T instead of concentrating on the rate of emergence. There are quite a few reasons for having taken this approach. First, as indicated in that section, we desired a verification of the particle-counting results obtained earlier. Second, the device used immediately removed the delta function appearing in the formulation of (7.8). The presence of such a generalized function should not be frightening or objectionable, however, since the entire procedure can be made rigorous using the theory of distributions. (We might mention that the delta function is, of course, present, in disguised form, in the physical description of the problem used directly in the particle-counting derivation of Section 2.) We might also note that from an experimental viewpoint the quantity $c\tilde{r}(x,T)$ would probably be more easily measured than its derivative, and is thus quite likely the more meaningful expression to seek.

However, in a sense this discussion is begging the important question: Just what is the invariant imbedding equation for the quantity $c(\partial\tilde{r}/\partial T)$? To answer this, let us start from Eq. (7.24) and the side conditions (7.13a, b). We define

$$r(x,T)=\frac{\partial\tilde{r}}{\partial T}(x,T) \tag{7.25}$$

Equation (7.24) may then be rewritten

$$\frac{\partial\tilde{r}}{\partial x}+\frac{2}{c}r(x,T)=\frac{\sigma}{c}+\sigma c\int_0^T r(x,t')\tilde{r}(x,T-t')\,dt'. \tag{7.26}$$

Assuming the validity of the operation we differentiate (7.26) with respect to T:

$$\frac{\partial r}{\partial x}+\frac{2}{c}\frac{\partial r}{\partial T}=\sigma c\int_0^T r(x,t')r(x,T-t')\,dt'+\sigma\, cr(x,T)\tilde{r}(x,0)$$

$$=\sigma c\int_0^T r(x,t')r(x,T-t')\,dt'. \tag{7.27}$$

In the last step we have made use of (7.13a). We must now find the side conditions on $r(x,T)$. Differentiating (7.13b) with respect to T yields

$$r(0,T)=0, \qquad T\geqslant 0. \tag{7.27a}$$

Similarly, differentiating (7.13a) with respect to x produces

$$\frac{\partial\tilde{r}}{\partial x}(x,0)=0.$$

Now when we evaluate (7.26) at $T=0$ we find

$$r(x,0)=\frac{\sigma}{2}\,. \tag{7.27b}$$

Equations (7.27) have been analyzed completely [1]. It should be noticed that the boundary condition (7.27b) is a rather surprising one. Had we tried to obtain equations (7.27) directly from Eqs. (7.8), we should probably not have expected a result like (7.27b) at all. Physically, it states that even at time zero, while no particles have actually emerged from the rod, they are *emerging* at a rate $(c\sigma/2)$. A cautious particle-counting analysis verifies that this phenomenon does indeed take place (see Problem 8). However, the result is a further indication that the *total* number of particles emergent up to a given time is a more natural and intuitive quantity with which to deal than is their rate of emergence.

Finally, it should be pointed out that the requirement $f=b=1$ has again been used here only for the sake of convenience. We leave more general parameters to the problem section. Obviously, the entire analysis may be carried over to the many-state case (the case of a matrix system) (see Problem 7).

Since, as indicated earlier in this volume, the reflection function seems to be the really physically significant quantity in all of these studies we shall not dwell on the transmission equations. Such matters are left to the reader (Problem 4).

5. TIME-DEPENDENT INPUT

Let us pose a slightly more complicated problem than that discussed in the preceding sections. We suppose the model is still that described by Eqs. (7.8) except that the input is not a delta function but is continuously distributed in time. Thus (7.6d) is replaced by

$$cv(x,t)=f(t), \qquad t\geqslant 0, \qquad f \text{ continuous.} \tag{7.6d$'$}$$

We now ask for the total number of particles emergent from the rod at $z=x$ from time zero to time T.

It is convenient to define in this case

$$\begin{aligned} c\hat{u}(z,T) &= \int_0^T cu(z,t')\,dt', \\ c\hat{v}(z,T) &= \int_0^T cv(z,t')\,dt', \\ \hat{f}(T) &= \int_0^T f(t')\,dt'. \end{aligned} \tag{7.28}$$

Straightforward calculations with Eqs. (7.8), (7.6c), (7.6e), and (7.6d′) yield

$$\begin{aligned} \frac{1}{c}\frac{\partial \hat{u}}{\partial T} + \frac{\partial \hat{u}}{\partial z} &= \sigma\hat{v}(z,T), \\ \frac{1}{c}\frac{\partial \hat{v}}{\partial T} - \frac{\partial \hat{v}}{\partial z} &= \sigma\hat{u}(z,T), \end{aligned} \tag{7.29}$$

$$c\hat{u}(0,T)=0, \qquad T\geqslant 0, \tag{7.29a}$$

$$c\hat{v}(x,T)=\hat{f}(T), \qquad T\geqslant 0, \tag{7.29b}$$

$$c\hat{u}(z,0)=c\hat{v}(z,0)=0, \qquad 0\leqslant z\leqslant x. \tag{7.29c}$$

Taking Laplace transforms with respect to T

$$L_T\{c\hat{u}(z,T)\}=\int_0^{\infty}e^{-sT}c\hat{u}(z,T)\,dT=\hat{U}(z,s),\ etc., \tag{7.30}$$

readily yields

$$\frac{d\hat{U}}{dz}=-\frac{s}{c}\hat{U}(z,s)+\sigma\hat{V}(z,s), \tag{7.31}$$

$$-\frac{d\hat{V}}{dz}=\sigma\hat{U}(z,s)-\frac{s}{c}\hat{V}(z,s),$$

$$\hat{U}(0,s)=0, \tag{7.31a}$$

$$\hat{V}(x,s)=\hat{F}(s)=L_T\{\hat{f}(T)\}. \tag{7.31b}$$

Equations (7.31) are precisely those satisfied by $\tilde{U}(z,s)$ and $\tilde{V}(z,s)$ as defined by (7.18), except that (7.18b′) has been replaced by (7.31b). The superposition principle implied by linearity immediately yields in the notation of (7.19).

$$\hat{U}(z,s)=\hat{F}(s)U(z,s),\qquad \hat{V}(z,s)=\hat{F}(s)V(z,s). \tag{7.32}$$

Thus

$$\hat{U}(x,s)=\{s\hat{F}(s)\}\tilde{R}(x,s) \tag{7.33}$$

upon using (7.21). Inverting in the usual way, we obtain

$$c\hat{u}(x,T)=\int_0^T\left[\frac{d}{dt}\int_0^t f(t')\,dt'\right]c\tilde{r}(x,T-t)\,dt$$

$$=\int_0^T f(t)c\tilde{r}(x,T-t)\,dt, \tag{7.34}$$

where $\tilde{r}$ is the function satisfying (7.24). Since the left side of Eq. (7.34) is the total number of particles reflected from the right end of the rod up to time T, $c\hat{u}$ is precisely the reflection function for the problem with the continuously distributed input. Actually, then, (7.34) is a statement of Duhamel's principle for the problem at hand. The function $c\tilde{r}$ is just the unit response function, a fact evident from its definition.

Obviously, a large collection of time-dependent problems may be studied in this way, including those with time-dependent inputs on the left as well as on the right, time-dependent sources distributed throughout the rod and varying with position as well as time, and so on. We leave some of these to the reader (Problem 9). In all such cases the solution may be considered an exercise in the technique of using the Laplace transform. Of course, if discontinuities in time in the sources are allowed care must be taken at various stages of the analysis.

6. THE TIME-DEPENDENT WAVE EQUATION

The problem we are about to discuss is simpler than the transport questions we have been considering. The time-dependent form of the wave equation studied in Chapter 6 is

$$\frac{\partial^2\psi}{\partial z^2} = k^2(z)\frac{\partial^2\psi}{\partial t^2}, \tag{7.35}$$

where $\psi = \psi(z,t)$. We now suppose a monochromatic wave impinges from the right onto the surface $z = x$ of a plane-parallel slab as considered in Chapter 6. This wave may be taken in the form

$$\psi(z,t) = e^{-i\alpha(z-x)}e^{i\omega t}. \tag{7.36}$$

Since the frequency ω of the wave cannot change in the reflection and transmission process, we replace $\psi(z,t)$ in Eq. (7.35) by

$$\psi(z,t) = \tilde{\psi}(z)e^{i\omega t}, \tag{7.37}$$

which reduces Eq. (7.35) to

$$\frac{d^2\tilde{\psi}}{dz^2} + \omega^2 k^2(z)\tilde{\psi}(z) = 0, \tag{7.38}$$

a familiar form.

In the region $z > x$, let $k(z)$ be the constant, k_2. Hence, we obtain from (7.35) and (7.36)

$$\alpha = k_2\omega. \tag{7.39}$$

Thus the wave form to the right of the slab must be

$$\psi(z,t) = \left\{e^{-ik_2\omega(z-x)} + \rho(x,\omega)e^{ik_2\omega(z-x)}\right\}e^{i\omega t}. \tag{7.40}$$

Here we have written $\rho(x,\omega)$ instead of simply $\rho(x)$, as in the previous chapter, in order to emphasize the dependence of the reflection coefficient on ω. Similarly, for $z < 0$ we easily determine that

$$\psi(z,t) = \tau(x,\omega)e^{-ik_1\omega z}e^{i\omega t}, \tag{7.41}$$

where k_1 is the (constant) wave number to the left of $z=0$.

The remainder of the analysis now parallels that of Chapter 6, or, more accurately, the results may be obtained directly from that chapter by simply making appropriate substitutions. Consequently we pursue the matter no farther.

It is interesting to note that the problem of a wave falling obliquely on a slab of the type we have been discussing has many points in common with the question we have just considered. We leave this analysis to the reader. Similarly the study of nonmonochromatic waves is left to the problem section (Problems 10–12).

7. THE DIFFUSION EQUATION

Let us now apply our methods to the classical diffusion equation. Since this is most frequently thought of as the equation describing the flow of heat in a body, we shall speak physically in those terms. However, it must be recalled that the same fundamental equation governs—at least approximately—a large variety of physical processes. In particular it is frequently used as the first, or crudest, approximation to the time-dependent transport equation to be discussed in Chapter 11 (see also [3, 4]). When the time-like variable is reinterpreted, the equation becomes the famous "neutron-age equation" of Fermi [4]. Thus the work of the next few sections is applicable to many physical phenomena.

We turn again to a slab geometry. Now assume that the slab is uniform in structure and capable of conducting heat. By proper choice of units the diffusivity may be chosen as unity. The temperature θ at any location z in the slab and at any time t is then a solution of the equation

$$\frac{\partial \theta}{\partial t}=\frac{\partial^2 \theta}{\partial z^2}. \tag{7.42}$$

Boundary and initial conditions are, of course, needed. Let us suppose that the temperature everywhere in the slab at time zero is itself zero, and that the left face of the slab is maintained at all times at this temperature. Finally, on the right surface of the slab, $z=x$, we specify a given flux of heat, this flux being a function of time. Mathematically,

$$\theta(z,0)=0, \qquad 0 \leqslant z \leqslant x, \tag{7.42a}$$

$$\theta(0,t)=0, \qquad 0 \leqslant t, \tag{7.42b}$$

$$-\frac{\partial \theta}{\partial z}(x,t)=f(t), \qquad 0 \leqslant t. \tag{7.42c}$$

What is the temperature $\theta(x,t)$ of the right face at any given time?

The problem is in a form which strongly suggests the use of the Laplace transform. Define

$$L_t(\theta)=\tilde{\theta}\,(z,s)=\int_0^{\infty} e^{-st}\theta(z,t)\,dt. \tag{7.43}$$

The equations resulting from (7.42) are

$$\frac{d^2\tilde{\theta}}{dz^2}-s\tilde{\theta}\,(z,s)=0, \tag{7.44}$$

$$\tilde{\theta}\,(0,s)=0, \tag{7.44a}$$

$$-\tilde{\theta}'(x,s)=\tilde{f}\,(s)=L_t(f). \tag{7.44b}$$

To obtain a more familiar structure we set

$$\tilde{\theta}\,(z,s)=\tilde{u}\,(z,s), \qquad \frac{d\tilde{\theta}}{dz}(z,s)=\tilde{v}\,(z,s), \tag{7.45}$$

and now have

$$\frac{d\tilde{u}}{dz}=\tilde{v}, \qquad -\frac{d\tilde{v}}{dz}=-s\tilde{u}, \tag{7.46}$$

$$\tilde{u}\,(0,s)=0, \tag{7.46a}$$

$$\tilde{v}\,(x,s)=-\tilde{f}\,(s). \tag{7.46b}$$

Equation (7.46b) is unfortunate; we should prefer that the right side be unity. A further transformation is called for:

$$-\tilde{U}(z,s)\tilde{f}\,(s)=\tilde{u}\,(z,s), \qquad -\tilde{V}(z,s)\tilde{f}\,(s)=\tilde{v}\,(z,s). \tag{7.47}$$

Now we have a set of equations that fit our previous theory nicely:

$$\frac{d\tilde{U}}{dz}=\tilde{V}, \qquad -\frac{d\tilde{V}}{dz}=-s\tilde{U}, \tag{7.48}$$

$$\tilde{U}(0,s)=0, \tag{7.48a}$$

$$\tilde{V}(x,s)=1. \tag{7.48b}$$

Equation (3.25) yields [we prefer at this stage *not* to call the function $\tilde{U}(x,s)$ by the usual name r]

$$\frac{d\tilde{U}}{dx}=1-s\tilde{U}^2(x,s). \tag{7.49}$$

The solution of (7.49) satisfying the side condition $\tilde{U}(0,s)=0$ is

$$\tilde{U}(x,s)=\frac{1}{\sqrt{s}}\tanh(x\sqrt{s}). \tag{7.50}$$

This result in turn yields

$$\tilde{u}(x,s)=-\frac{\tilde{f}(s)}{\sqrt{s}}\tanh(x\sqrt{s}). \tag{7.51}$$

Thus the temperature at the right side of the slab at any time t is given by the convolution integral

$$\theta(x,t)=L_s^{-1}[\tilde{u}(x,s)]=-\int_0^t dt' f(t-t')L_s^{-1}\left[\frac{\tanh(x\sqrt{s})}{\sqrt{s}}\right]_{t'}. \tag{7.52}$$

Although the invariant imbedding analysis is now complete, it is certainly appropriate to pursue the problem a bit farther and discuss the inverse transform that appears in (7.52). According to [5],

$$L_t\left[\theta_1\left(\frac{1}{2},t\right)\right]=-\frac{1}{\sqrt{s}}\tanh\sqrt{s}, \tag{7.53}$$

where θ_1 is the theta-function of index one. Now if

$$g(s)=L_t[G(t)], \tag{7.54}$$

then

$$g(ks)=\frac{1}{k}L_t\left[G\left(\frac{t}{k}\right)\right]. \tag{7.55}$$

Therefore

$$L_s^{-1}\left[\frac{1}{\sqrt{s}}\tanh(x\sqrt{s})\right]=-\frac{1}{x}\theta_1\left(\frac{1}{2},\frac{t}{x^2}\right), \tag{7.56}$$

which allows the expression given by (7.52) to be rewritten

$$\theta(x,t)=\frac{1}{x}\int_0^t \theta_1\left(\frac{1}{2},\frac{t'}{x^2}\right)f(t-t')\,dt'. \tag{7.57}$$

The properties of the theta function θ_1 are well known and it is not pertinent to go into them here (see, for example, [6]). It is essential to emphasize that the result obtained is by no means new; indeed it is quite classical. The approach via invariant imbedding, is an interesting alternative method.

8. SOME COMMENTS ON THE PREVIOUS SECTION

Section 7 involved a rather large number of transformations of the dependent variable in order to obtain equations to which the standard imbedding equations could be directly applied. It is appropriate to ask if some of these could have been avoided. Another way to pose this question is to ask the physical meaning of Eqs. (7.48). Suppose we again allow ourselves the "luxury" of the delta-function and replace the function $f(t)$ in Eq. (7.42c) by $\delta(t)$. Physically, the slab has received an instantaneous unit heat flux at time $t=0$ on its right surface and there is no further input. The calculations of the last section can be repeated with the results

$$\frac{d\tilde{u}}{dz}=\tilde{v}, \qquad -\frac{d\tilde{v}}{dz}=-s\tilde{u}, \tag{7.58}$$

$$\tilde{u}(0,s)=0, \tag{7.58a}$$

$$\tilde{v}(x,s)=-1. \tag{7.58b}$$

We easily obtain the result that

$$\tilde{u}(x,s)=\frac{-1}{\sqrt{s}}\tanh(x\sqrt{s}\,), \tag{7.59}$$

which can be inverted as in the previous section. To be consistent with our previous analyses it is reasonable to think of this as the "reflection function" for the problem and write

$$\tilde{r}(x,t)=-L_s^{-1}\left[\frac{1}{\sqrt{s}}\tanh(x\sqrt{s}\,)\right]. \tag{7.60}$$

The final result (7.57) is thus once again a statement of Duhamel's theorem.

Very roughly speaking, most of the problems with which we are dealing can easily be handled by finding the response of the linear system in question to some sort of a unit impulse, considering this response as a *generalized* reflection (or transmission) function and then applying standard methods of classical analysis to obtain the response of the system to a more general input.

9. A CRITIQUE OF SECTIONS 7 AND 8

As noted in the previous section, the result we have obtained is classical; in fact, there are much easier (or at least more familiar) ways of finding it than the one we have chosen to use. What then are the advantages—if any—of the invariant imbedding method when applied to such problems?

At least one advantage may be noted if one examines the structure of Eq. (7.44). There are two fundamental solutions to this equation, $e^{z\sqrt{s}}$ and $e^{-z\sqrt{s}}$. Suppose one did not recognize this fact and attempted to solve (7.44) on a computer by first finding these fundamental solutions and then using superposition to satisfy any boundary conditions. For s large or x large or both, the positive exponential would rather quickly "swamp" the calculation. One could, of course, turn to other numerical methods designed especially to handle two-point boundary value problems. Although these may well avoid the difficulty mentioned, they often involve considerably more computational work. Certainly any "shooting" type method used in a direct fashion would almost surely encounter trouble.

On the other hand, if one were to attack Eq. (7.49) numerically he would find little difficulty regardless of the size of either s or x. Moreover, the problem being solved is an initial value problem—no superposition is required.

Although these considerations are certainly academic in the problem under study, they become quite important once the assumption that the diffusivity is a constant is abandoned. If one assumes this physical quantity to depend on the position, z, then an explicit analytic solution of the equation which replaces (7.44) is very likely impossible. Since the diffusivity is a positive function the analog of (7.44) will generally be such that its fundamental solutions have behaviors somewhat like the growing and decaying exponentials encountered in the easy case we have studied. The analog of (7.49) will usually continue to have a reasonable behavior that will not lead to excessive numerical problems.

Another definite advantage is possessed by the imbedding method in the more realistic case. If one wishes to solve the problem for a variety of thicknesses (x values) he need only continue the integration of the analog of (7.49) out to the largest x needed, recording the values of $\tilde{U}$ at the various intermediate x values of interest. Many of the effective methods for numerically integrating the corresponding two-point boundary value problem are of such a nature that a completely new calculation must be done each time the thickness x is changed.

Finally, the imbedding method yields precisely the quantity we were seeking—namely, the temperature on the right face of the slab. At least in its direct application, the method does not produce any other information that might be extraneous to our immediate interests. The classical approach would produce a great deal of data about internal temperatures as well.

We may, however, turn the argument around and ask: Suppose the internal temperatures *are* of value. Can the imbedding approach still be used? To make the problem even more challenging, let us assume that the temperature itself is assigned at both left and right faces—a common physical situation. We turn to this entirely new question in the next section.

10. ANOTHER DIFFUSION PROBLEM

We pose the following problem:

$$d(z)\frac{\partial\theta}{\partial t}=\frac{\partial^2\theta}{\partial z^2},\qquad 0\leqslant z\leqslant x,\qquad t>0, \tag{7.61}$$

$$\theta(z,0)=0,\qquad 0<z<x, \tag{7.61a}$$

$$\theta(0,t)=g_1(t),\qquad t\geqslant 0, \tag{7.61b}$$

$$\theta(x,t)=g_2(t),\qquad t\geqslant 0. \tag{7.61c}$$

We seek the temperature θ for $0\leqslant z\leqslant x, t\geqslant 0$. In (7.61) the function $d(z)$ is the reciprocal of the diffusivity. It is positive and at least piecewise continuous.

We proceed to take the Laplace transform of (7.61) and use the same functions $\tilde{u}$ and $\tilde{v}$ as introduced in Eq. (7.45). The resulting problem is

$$\frac{d\tilde{u}}{dz}=\tilde{v},\qquad -\frac{d\tilde{v}}{dz}=-sd(z)\tilde{u}, \tag{7.62}$$

$$\tilde{u}(0,s)=\tilde{g}_1(s), \tag{7.62a}$$

$$\tilde{u}(x,s)=\tilde{g}_2(s). \tag{7.62b}$$

This set of conditions suggests turning to Chapter 3. However, the problem is not of the precise form dealt with in Chapter 3; both boundary conditions are on the u function instead of one being on the u and the other on the v function. Nevertheless, the functional equations of that chapter are applicable.

It must be recalled that the basic Eqs. (3.3) contain a tremendous amount of information, and that the numerous results of Chapter 3 were derived directly from them. It is not surprising, then, that proper manipulation of these equations can produce a result useful for our present consideration. Such a formula is found in Problem 11 of Chapter 3:

$$u(z)=\big[u(z_1)\{t_l(z_1,z)r_r(z_1,z_2)-t_l(z_1,z_2)t_r(z,z_2)r_r(z_1,z)\}$$
$$+u(z_2)t_r(z,z_2)r_r(z_1,z)\big]\big[r_r(z_1,z_2)\{1-r_r(z_1,z)r_l(z,z_2)\}\big]^{-1}. \tag{7.63}$$

Making the obvious identification of variables, we conclude for the problem (7.62):

$$\tilde{u}(z,s)=\big[\tilde{g}_1(s)\{t_l(0,z,s)r_r(0,x,s)-t_l(0,x,s)t_r(z,x,s)r_r(0,z,s)\}$$
$$+\tilde{g}_2(s)t_r(z,x,s)r_r(0,z,s)\big]\big[r_r(0,x,s)\{1-r_r(0,z,s)r_l(z,x,s)\}\big]^{-1}. \tag{7.64}$$

Obviously, it is necessary to calculate several reflection and transmission functions, but the fact that these all involve initial value problems considerably simplifies the computation. For example,

$$\frac{dr_r}{dz}=1-sd(z)r_r^2(0,z,s), \qquad r_r(0,0,s)=0, \qquad s\geqslant 0. \tag{7.65}$$

The equations for the other r and t functions may be found among the problems at the end of Chapter 3. Since $d(z)$ is a relatively arbitrary function, numerical methods must almost certainly be employed even at this stage, and then the Laplace inverse must still be taken—presumably also numerically. We do not wish to underestimate some of the difficulties involved, which may indeed by quite formidable. However, we hope that we have made clear that the imbedding method may still prove of value in cases much more complicated than that considered in Section 7. It should also be recalled that information about the original heat flow problem may in general be obtained for small or large times by examining the Laplace transform for large or small s values, respectively. For this type of investigation full scale inversion of $\tilde{u}(z,s)$ is not required.

11. A FINAL DIFFUSION PROBLEM

It seems appropriate to look at one further diffusion problem of a type that frequently arises in practice. The example we choose to examine has internal heat sources, with Newton's law of cooling satisfied at each slab face. Typical equations are

$$d(z)\frac{\partial\theta}{\partial t}=\frac{\partial^2\theta}{\partial z^2}+f(z,t), \tag{7.66}$$

$$\theta(z,0)=0, \qquad 0\leqslant z\leqslant x, \tag{7.66a}$$

$$-\frac{\partial\theta}{\partial z}(0,t)+h_1\theta(0,t)=0, \qquad t>0, \qquad h_1>0, \tag{7.66b}$$

$$+\frac{\partial\theta}{\partial z}(x,t)+h_2\theta(x,t)=0, \qquad t>0, \qquad h_2>0. \tag{7.66c}$$

We turn at once to the Laplace transform:

$$\frac{d^2\tilde{\theta}}{dz^2}-s\,d(z)\tilde{\theta}\,(z,s)=-\tilde{f}\,(z,s), \tag{7.67}$$

$$-\frac{d\tilde{\theta}}{dz}(0,s)+h_1\tilde{\theta}\,(0,s)=0, \tag{7.67a}$$

$$\frac{d\tilde{\theta}}{dz}(x,s)+h_2\tilde{\theta}\,(x,s)=0. \tag{7.67b}$$

The boundary conditions at $z=0$ and $z=x$ are obviously quite different from any that we have encountered in this chapter, but they do suggest Eqs. (6.64). We therefore try the same kind of trick employed in dealing with the wave equation, defining

$$\tilde{u}\,(z,s)=-\frac{d\tilde{\theta}}{dz}(z,s)+h_1\tilde{\theta}\,(z,s), \tag{7.68a}$$

$$\tilde{v}\,(z,s)=\frac{d\tilde{\theta}}{dz}(z,s)+h_2\tilde{\theta}\,(z,s)+1. \tag{7.68b}$$

Equations (7.67) now become

$$\frac{d\tilde{u}\,(z,s)}{dz}=H^{-1}[(-s\,d(z)-h_1h_2)\tilde{u}$$
$$+(-s\,d(z)+h_1^2)\tilde{v}\,]-\tilde{f}\,(z,s)+H^{-1}(h_1^2-s\,d(z)),$$

$$-\frac{d\tilde{v}}{dz}(z,s)=H^{-1}[(-s\,d(z)+h_2^2)\tilde{u}$$
$$+(-s\,d(z)-h_1h_2)\tilde{v}\,]+\tilde{f}\,(z,s)+H^{-1}(h_1h_2+s\,d(z)), \tag{7.69}$$

where we have written $H=h_1+h_2$. Also

$$\tilde{u}(0,s)=0, \tag{7.69a}$$

$$\tilde{v}(x,s)=1. \tag{7.69b}$$

Since Eq. (7.69) is now in precisely the form (2.47) we may employ the methods developed in Chapter 2 for its solution. Of course, it must be recalled that the formidable task of taking the Laplace inverse still remains. The temperature θ and the temperature gradient $\partial\theta/\partial z$ may be obtained from (7.68) after inversion. We note, incidentally, that the Dirac delta function has been introduced into this problem in a rather concealed fashion—namely, the inverse of $\tilde{v}(z,s)$ will contain the delta function because of the presence of the number one included in the definition of $\tilde{v}$. In the light of the remarks made in Section 8 this should no longer prove shocking or even surprising.

Again, we carry this example no further; the reader is referred to the problems.

12. SUMMARY

This chapter has been devoted to the development of the invariant imbedding method for time-dependent problems. We have recognized that the introduction of the time variable very considerably complicates the analysis. Early attempts, confined to the study of simple transport problems, were based on the relatively primitive particle-counting technique. Results were difficult to obtain and frequently contained errors. With the idea of defining the reflection function in such a manner that the source particle could be introduced at an arbitrary time instead of just at time zero, considerable progress was made, but anything approaching a general theory was still most elusive.

After the general ideas contained in the first three chapters of this book were worked out, it became clear that the classical Laplace transform device could be profitably used on many problems to reduce them to a more tractable form. We have carried through several examples using this idea, leaving many of the details to the problem section.

It must also be noted, however, that the aforementioned "primitive" particle-counting technique has an advantage not possessed by the Laplace transform scheme. Namely, it is possible to handle problems in transport theory whose physical parameters depend directly on time; this is not in general possible with the transform method (see Problems 1, 2, and 5). It is reasonably certain that particle-counting methods may be applied to diffusion type problems to handle cases in which the physical parameters (such as d, the diffusivity) are time-dependent. The fact that diffusion theory is, from one viewpoint, a very crude approximation to transport theory [3,4] makes this seem an even more likely possibility. Little work has been done in this direction, however.

In any event, it must be admitted that the time-dependent invariant imbedding problems are much less attractive—at least from a computational viewpoint—than are the time-independent cases. However, it is well known that the introduction of time variation in even classically posed problems vastly increases their numerical complexity. From the strictly mathematical viewpoint, we notice that the time-dependent invariant imbedding equations can have a very interesting and unusual form and structure [see, for example, Eq. (7.11)]. The analysis of the existence of a solution, its uniqueness, its asymptotic properties, and so on, provide the mathematician with a considerable challenge (see [1, 2]).

PROBLEMS

1. By particle-counting methods derive the analog of Eq. (7.11) for the case in which f, b, and σ all depend on both z and t. Specialize to the case in which these functions depend only upon z and obtain the analog of (7.27).

2. Consider a time-dependent transport problem in which the medium knows its right from its left. That is, f and b will not be the same for both right- and left-moving particles. Derive the classical Boltzmann equations for this case and the corresponding invariant imbedding equation for $\tilde{r}$. Assume originally that f, b, and σ are all dependent upon t as well as upon z, then specialize.

3. Given the system

$$\frac{\partial u}{\partial z}+\frac{1}{c}\frac{\partial u}{\partial t}=a(z,t)u(z,t)+b(z,t)v(z,t),$$

$$-\frac{\partial v}{\partial z}+\frac{1}{c}\frac{\partial v}{\partial t}=c(z,t)u(z,t)+d(z,t)v(z,t),$$

with $u(z,0)=v(z,0)=0$, $u(0,t)=0$, $v(x,t)=\delta(t-\tau)$, where $a(z,t)$, $b(z,t)$, etc., are reasonably arbitrary functions. By interpreting this as a "generalized" transport problem, obtain corresponding invariant imbedding equation for $\tilde{r}$ by particle-counting.

4. In the first three problems derive equations satisfied by the transmission functions.

5. Study the time dependent problem in which the coefficient functions (f, b, etc.) are *independent* of t. Show that the Laplace transform method of the text can be used to handle all such problems and the equations for reflection and transmission functions can be easily found. Check your results with those obtained in the preceding four problems. Observe that the analog of the function $\tilde{r}$ of (7.12) may always be utilized.

6. Does the "optical depth" transformation (see Problem 4 of Chapter 1) work in the time-dependent case? Discuss this from both the physical and mathematical viewpoints.

7. Develop a time-dependent theory for the many-state case, using both the classical and invariant imbedding approaches.

8. Verify by particle-counting that the phenomenon involving (7.27b) actually can be understood physically. Are there other problems in mathematical physics where this sort of anomaly occurs?

9. In the text the source-free transport problem has been considered. Study the transport problem with internal sources s^+ and s^- that are dependent on both z and t. Show that the Laplace transform technique is applicable even when the source is time-dependent provided the functions f, b, and so on, are time independent. In the case that these latter functions do depend on time try to use particle-counting techniques to obtain invariant imbedding equations.

10. Resolve the case of the time-dependent wave equation in which a plane wave falls obliquely on the medium.

11. Study the wave equation problem when the input wave is not one of a single frequency but is continuously dependent upon frequency. Obtain equations for the reflection and transmission functions.

12. Combine the ideas included in Problems 10 and 11.

13. Find the invariant imbedding equations for the problem discussed in Section 11. By choosing special values for $d(z)$, $f(z,t)$, h_1, and h_2 verify that your results reduce to some of those obtained in earlier problems.

14. Consider the diffusion equation $u_t = u_{xx}$. Define a new function v by the relationship $u_x = vu$. (Observe the similarity between this transformation and those used in the early chapters of this book.) Show that $u_{xx} = v_x u + v u_x$, $u_{xt} = v_t u + v u_t$, and also $u_{xt} = v_{xx} u + 2 v_x u_x + v u_t$. Hence demonstrate that v satisfies the nonlinear equation of Burger, $v_t = v_{xx} + 2 v v_x$.

15. Start with the equation $u_t = u_{xxx} + a_1 u_{xx} + a_2 u_x + a_3 u$. Now set $(u_{xx} + b_1 u_x + b_2 u) = v(c_1 u_{xx} + c_2 u_x + c_3 u)$. Try to choose the functions b_j and c_j so as to obtain the Kortweg–de Vries equation $v_t = v_{xxx} + v v_x$. If this is not possible, try to "come close" to this equation. (See P. D. Lax, *Integrals of Nonlinear Equations of Evolution*, Courant Institute, New York University, 1968, NYO-1480-87.)

16. Given the heat conduction problem $u_t = u_{xx}$, $t > 0$, $0 < x < 1$, $u(0,t) = u(1,t) = 0$, $u(x,0) = g(x)$. Obtain an invariant imbedding formulation of the problem and solve. Consider the more complicated equation $d(x) u_t = u_{xx}$ subject to the same side conditions.

17. Let $f(s)$ be the Laplace transform of $F(t)$, $f(s) = \int_0^\infty e^{-st} F(t)\,dt$. Demonstrate that a "small" change in $F(t)$ results in only a "small" change in $f(s)$. Now show by example that a "small" change in $f(s)$ may result in a "large" change in $F(t)$. Discuss the unpleasant implications of this observation, especially from the viewpoint of trying to obtain inverse Laplace transforms numerically.

18.* Consider Eqs. (7.8) of the text. Formally, as $c \to \infty$ these equations go into $\tilde{u}' = \sigma \tilde{v}$, $-\tilde{v}' = \sigma \tilde{u}$. Are the solutions of this new set of ordinary differential equations the limits of the solutions to Eqs. (7.8)? If so, what are the appropriate boundary conditions?

19.* The formal limit of Eq. (7.13) as $c \to \infty$ and $\sigma c \to 1$ is

$$\frac{\partial \tilde{r}}{\partial x} = \int_0^T \frac{\partial \tilde{r}}{\partial t'}(x,t')\tilde{r}(x,T-t')\,dt'.$$

Does the solution of (7.13) approach that of this equation provided appropriate boundary conditions are imposed? (See R. Bellman, R. Kalaba, and G. M. Wing, "Invariant Imbedding and Neutron Transport. V. Diffusion as a Limiting Case," *J. Math. Mech.* **9**, 1960, 933–944.)

REFERENCES

1. G. M. Wing, "Solution of a Time Dependent, One Dimensional Neutron Transport Problem," *J. Math. Mech.* 7, 1958, 757–766.
2. G. M. Wing, "Analysis of a Problem of Neutron Transport in a Changing Medium," *J. Math. Mech.* **11**, 1962, 21–34.
3. A. M. Weinberg and E. P. Wigner, *The Physical Theory of Neutron Chain Reactors*, University of Chicago Press, Chicago, 1958.
4. S. Glasstone and M. C. Edlund, *The Elements of Nuclear Reactor Theory*, Van Nostrand, New York, 1952.
5. H. Bateman, in *Tables of Integral Transforms* (A. Erdelyi, Ed.), McGraw Hill, New York, 1952.
6. H. Bateman, *Higher Transcendental Functions*, McGraw Hill, New York, 1952.

8

THE CALCULATION OF EIGENVALUES FOR STURM-LIOUVILLE TYPE SYSTEMS

1. INTRODUCTION

In the previous chapters we have developed the method of invariant imbedding with emphasis on its application to problems that arise from the consideration of various physical processes. Despite the fact that very considerable impetus for this development has been provided by numerical considerations, we have tended to deemphasize that aspect of the technique. It is appropriate at this point to include more on some calculational matters than was done in earlier chapters.

In this chapter, then, we quite frankly and unhesitatingly turn to numerical considerations. Our purpose here is to show how the imbedding method can be applied in a direct and simple fashion to the calculation of eigenvalues (and eigenlengths) for equations of Sturm–Liouville type. As a matter of fact, the method is so easy that it is surprising that it has not been used before in its present form. As we shall point out, ideas quite similar have been employed previously, but in most cases they turn out to be less efficient and less accurate. Our presentation follows quite closely the work of [1].

2. EIGENLENGTHS FOR TRANSPORT-LIKE EQUATIONS IN ONE DIMENSION

We return to a familiar system

$$\frac{du}{dz}=a(z)u(z)+b(z)v(z), \tag{8.1a}$$

$$-\frac{dv}{dz}=c(z)u(z)+d(z)v(z). \tag{8.1b}$$

If this is a model for an actual transport problem, and if the medium is multiplying—that is, if on the average more than one particle emerges from a collision—then we expect that the system can go critical. From one physical viewpoint this means that if the system extends from zero to x_{cr}, the critical length, and if any particles are injected into the system, then no solution (u,v) exists. From another viewpoint—again physical—a system of exactly this size will support a particle population (u,v) without any outside sources. The shape of this population is determined by the parameters involved; the size of the population cannot be predetermined because (8.1) is linear and we are now actually considering boundary conditions

$$u(0)=0, \tag{8.1c}$$

$$v(x_{cr})=0. \tag{8.1d}$$

In more mathematical terms we are really discussing a very special case of the Fredholm alternative. (See [2], also Problem 1). Provided the coefficients $a(z)$, $b(z)$, $c(z)$, and $d(z)$ are at all reasonably behaved, the same phenomenon will quite likely occur for other x values. That is, the system (8.1) will have a nontrivial solution for $x_{cr}=x_1<x_2<x_3\cdots$. The physical meaning is lost for these larger eigenlengths, but the mathematical phenomenon persists. We have actually encountered this situation in earlier work (see Chapter 1, Section 3). We shall now make good use of this observation.

3. THE CALCULATION OF EIGENLENGTHS

Having motivated our goals via a transport-like problem, let us introduce a suitable reflection function. Following the lead of Chapter 3 we set

$$u(z)=r(z)v(z). \tag{8.2}$$

Henceforth let us suppose that the coefficients $a(z)$, $b(z)$, $c(z)$, and $d(z)$ are such that the usual existence and uniqueness theorems hold for *initial value problems* associated with (8.1a, b) on the interval $0 \leqslant z \leqslant \tilde{x}$. We shall be interested only in nontrivial solutions. It is at once clear that if (8.1c) is to hold then $v(0) \neq 0$, for otherwise $u(z) = v(z) = 0$ for all z, $0 \leqslant z \leqslant \tilde{x}$. It follows from well-known theory that there is some interval to the right of zero on which $v(z) \neq 0$. We seek $x_{cr} = x_1$, the first zero of v.

Now if we substitute (8.2) into (8.1a, b) and do a bit of manipulating we obtain [see Problem 2 and Eq. (3.25)].

$$r'(z) = b(z) + (a(z) + d(z))r(z) + c(z)r^2(z). \tag{8.3}$$

In order for (8.1c) to be valid we must require

$$r(0) = 0. \tag{8.3a}$$

Obviously, (8.3) is the equation for the reflection function for the problem, a fact that should by now not be the least bit surprising. Moreover, the function $r(z)$ is defined by Eq. (8.3) just so long as $v(z) \neq 0$. If, however, $v(z)$ has a zero then r cannot be defined at that point, for if it were, then by Eq. (8.2) u would have a zero at the same point and uniqueness again rules this out. Hence, the solution of (8.3) fails to exist at such a point—which is, of course, just $x_{cr} = x_1$.

Thus far, we have accomplished little that is new except that we have put some of the matters with which we have so often dealt in this book on a somewhat more rigorous basis. However, we must now inquire into a method of finding x_2, assuming such a point exists. And, as has been pointed out, physical intuition now breaks down.

To accomplish our task we introduce a function $s(z)$ by formally writing

$$s(z) = \frac{1}{r(z)} \tag{8.4}$$

so that

$$v(z) = s(z)u(z). \tag{8.5}$$

Now, clearly $s(z)$ is not well defined at $z = 0$. Indeed, we make no effort to define $s(z)$ there. Rather, we choose a point z_1', $0 < z_1' < x_1$ at which the r function is well behaved and nonzero. Substituting (8.5) into (8.1a, b) yields

$$-s'(z) = c(z) + [a(z) + d(z)]s(z) + b(z)s^2(z), \tag{8.6}$$

and this may be integrated starting forward at $z = z_1'$, subject to the condition

$$s(z_1') = \frac{1}{r(z_1')}. \tag{8.6a}$$

We knew in advance that $s(z)$ would also satisfy an equation of Riccati type. An analysis of the derivation of equation (8.6) reveals that this Riccati equation is valid provided $u(z) \neq 0$. It will do no harm to suppose that no zero of u exists between z_1' and x_1. (If one did, then we could simply make another choice of z_1'.) Uniqueness assures us that $u(x_1) \neq 0$. Thus Eq. (8.6) may be integrated to the right of x_1. We note in particular that $s(x_1)=0$. As the integration approaches the next zero of u, call it $\hat{x}_1$, the function $s(z)$ will grow in absolute value, for it cannot exist at $\hat{x}_1$, again by the uniqueness argument.

To avoid this difficulty, we choose a point z_1'', $x_1 < z_1'' < \hat{x}_1$, and return to the function $r(z)$ defined by (8.2). It still obeys Eq. (8.3) but now satisfies the initial condition

$$r(z_1'') = \frac{1}{s(z_1'')} . \tag{8.7}$$

The Riccati equation for r may now be integrated forward to "blow-up," and this will be precisely the point x_2. Of course, it is more reasonable in practice to switch over again to the s function before x_2 is reached. Since the s equation (8.6) is well behaved at x_2 (indeed $s(x_2)=0$), the integration may be carried beyond that point with no difficulty.

The scheme is now clear. One starts with the r function, integrates until it begins to become large (in absolute value), switches to the s function until it begins to misbehave, returns to the r function, and so on. The zeros of the s function are precisely the points $x_1, x_2, x_3, \ldots$, that we were seeking—the eigenlengths of the problem defined by the entire system [Eqs. (8.1a, b, c, d)]. Figure 8.1 indicates how the various x and z points that enter into the calculation are arranged.

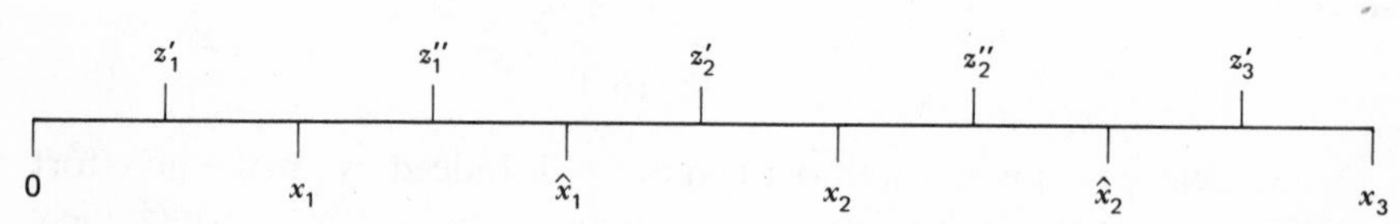

Figure 8.1.

The astute reader will have noticed that as a by-product we have also solved the problem defined by Eqs. (8.1a, b, c) and the condition

$$u(\tilde{z}_j)=0, \qquad j=1,2,\ldots. \tag{8.1d$'$}$$

4. SOME GENERALIZATIONS

We have noted at the end of the previous section that, incidental to the solution of the problem under consideration, we have solved another problem—namely, one in which the value of u is assigned as zero at both endpoints. This suggests at once that still more complicated boundary conditions may be investigated by the method we have devised.

Consider the set (8.1a, b) with side conditions

$$a_1u(0)+a_2v(0)=0, \tag{8.8a}$$

$$b_1u(x)+b_2v(x)=0, \tag{8.8b}$$

where a_1, a_2, b_1, b_2 are more or less arbitrary constants (see Problem 4). We seek those values of x, $x_1<x_2<x_3<\cdots$, such that (8.1a, b) and (8.8) are satisfied. The substitution (8.2) then yields

$$r(0)=-\frac{a_2}{a_1}, \tag{8.9a}$$

and

$$r(x)=-\frac{b_2}{b_1}, \tag{8.9b}$$

while the s substitution gives

$$s(0)=-\frac{a_1}{a_2} \tag{8.10a}$$

$$s(x)=-\frac{b_1}{b_2}. \tag{8.10b}$$

One may now use exactly the scheme outlined in the previous section, simply replacing the earlier conditions that r and s be zero by the conditions given by (8.9) or (8.10). In the event that one of the expressions on the right-hand side of those equations involves a formal division by zero, say (8.9a), then the corresponding one—in this case (8.10a)—may be used to start the integration.

Again, it should be noted that any nonsingular transformation, of the type

$$\begin{aligned}\tilde{u}(z)&=a_{11}u(z)+a_{12}v(z),\\ \tilde{v}(z)&=a_{21}u(z)+a_{22}v(z).\end{aligned} \tag{8.11}$$

converts the equations (8.1a, b) into a system of the same kind, and the same remark may be made concerning the side conditions (8.8). In fact, the transformation (8.11) with $a_{11}=a_1$, $a_{12}=a_2$, $a_{21}=b_1$, and $a_{22}=b_2$ is a particularly obvious and appealing one (see Problem 4). In any event, substitutions of the form (8.11) may be obtainable to simplify the entire problem, and such substitutions should be considered before numerical integration is begun.

Finally, it should be noted that any two-point boundary value problem, of the type

$$y''+a(z)y'+b(z)y=0, \tag{8.12a}$$

$$a_3y(0)+a_4y'(0)=0, \tag{8.12b}$$

$$b_3y(x)+b_4y'(x)=0, \tag{8.12c}$$

can be put in the form of the systems we have been investigating by means of the transformation

$$u(z)=y(z), \tag{8.13a}$$

$$u'(z)=v(z). \tag{8.13b}$$

This observation leads us naturally to the eigenvalue problem for Sturm–Liouville systems.

5. RESULTS FOR STURM–LIOUVILLE SYSTEMS

We now examine the Sturm–Liouville system

$$\frac{d}{dz}\left\{k(z,\lambda)\frac{du}{dz}\right\}+g(z,\lambda)u(z)=0, \tag{8.14a}$$

$$\begin{aligned} a_1(\lambda)u(0)+b_1(\lambda)u'(0)&=0,\\ c_1(\lambda)u(x)+d_1(\lambda)u'(x)&=0. \end{aligned} \tag{8.14b}$$

We consider x fixed and λ a parameter. It is well known [3] that for suitable functions k and g, and appropriate $a_1(\lambda)$, $b_1(\lambda)$, and so on, there is a discrete set of eigenvalues λ for the system (8.14). One of the conditions that we particularly need is that $k(z,\lambda)\neq 0$ on $0\leqslant z\leqslant x$. Then the substitution

$$\frac{du}{dz}=\frac{1}{k(z,\lambda)}v(z), \tag{8.15a}$$

is meaningful and leads to

$$-\frac{dv}{dz}=g(z,\lambda)u(z). \tag{8.15b}$$

We now have a special case of Eqs. (8.1a, b) subject to conditions like (8.8). Clearly, a particular value of λ is an eigenvalue if x is an eigenlength in the sense of Section 3. The corresponding r and s functions are solutions of

$$\frac{dr}{dz}=\frac{1}{k(z,\lambda)}+g(z,\lambda)r^2(z) \tag{8.16a}$$

and

$$-\frac{ds}{dz}=g(z,\lambda)+\frac{1}{k(z,\lambda)}s^2(z). \tag{8.16b}$$

The boundary conditions take the form

$$r(0)=\frac{-b_1(\lambda)}{k(0,\lambda)a_1(\lambda)}, \tag{8.17a}$$

$$r(x)=\frac{-d_1(\lambda)}{k(x,\lambda)c_1(\lambda)}. \tag{8.17b}$$

Of course, Eqs. (8.17) are replaced by their equivalents involving the s function in case either of these last expressions is formally infinite.

The numerical scheme for computing eigenvalues is as follows. Guess a value for λ. Now use the method outlined in the previous sections to find the corresponding eigenlengths. If one of these eigenlengths is precisely the x occurring in (8.14b) then the guess is indeed a correct eigenvalue. If not, then repeat the process with another λ, and so on. In practice, one actually computes curves of λ versus eigenlength, and then picks off the correct eigenvalues by selecting the desired value of x. This will become clearer in the section in which specific examples are studied. (See especially Figure 8.2.) At the moment the method may appear to be quite time-consuming and inaccurate; actually, the opposite is true in the many cases studied. One of the reasons for the relative accuracy as compared to some other popular methods is indicated in the next section.

A theoretical advantage to our method may also be seen by examining

Eqs. (8.18)

$$r(z)=\frac{u(z)}{v(z)}=\frac{u(z)}{k(z,\lambda)u'(z)}, \tag{8.18a}$$

$$s(z)=\frac{k(z,\lambda)u'(z)}{u(z)}. \tag{8.18b}$$

When only the eigenvalue of the problem is of interest the specific values of $u(z)$ and $u'(z)$ are of no particular importance; the above equations indicate that only their ratios play a genuine role in our algorithm. The step size used in a numerical calculation can be adjusted according to this ratio. A more direct approach, based on straightforward integration of the original system (8.14), would be sensitive to the behavior of $u(z)$ and $u'(z)$ individually and the integration step size would be dictated by the poorer of these.

6. CONNECTION WITH THE PRÜFER TRANSFORMATION

Recently the classical Prüfer transformation has been used by several authors to analyze the kind of problem we have been discussing (see, e.g., [4]). Since the method is quite closely related to ours, some comparison is called for. Define the functions $\rho(z)$ and $\theta(z)$ by the relations

$$u(z)=\rho(z)\sin\theta(z), \tag{8.19a}$$

$$k(z,\lambda)u'(z)=\rho(z)\cos\theta(z). \tag{8.19b}$$

Substitution in (8.14a) leads readily to the equation

$$\theta'(z)=\frac{1}{k(z,\lambda)}\cos^2\theta(z)+g(z,\lambda)\sin^2\theta(z). \tag{8.20}$$

Since

$$\tan\theta(z)=\frac{u(z)}{k(z,\lambda)u'(z)}, \tag{8.21}$$

the boundary conditions (8.14b) can be expressed in the form

$$\theta(0)=w_1, \tag{8.21a}$$

$$\theta(x)=w_2+n\pi, \qquad n=0,1,2,\ldots, \tag{8.21b}$$

where

$$\tan w_1 = -\frac{b_1(\lambda)}{k(0,\lambda)a_1(\lambda)}, \tag{8.22a}$$

$$\tan w_2 = -\frac{d_1(\lambda)}{k(x,\lambda)c_1(\lambda)}. \tag{8.22b}$$

Also, from (8.18) we find

$$r(z) = \tan\theta(z), \tag{8.23a}$$

$$s(z) = \cot\theta(z). \tag{8.23b}$$

Finally, we observe that the r equation (8.16a) can be obtained from (8.20) by division by $\cos^2\theta(z)$; the s equation (8.16b) can be similarly obtained by division by $\sin^2\theta(z)$.

The relationship between the two approaches is quite evident. However, the imbedding scheme seems to have two distinct advantages, at least. First, as we have seen in earlier chapters, the invariant imbedding method readily generalizes to handle systems of equations—or, equivalently, equations of high order. A corresponding generalization of the Prüfer transformation is by no means clear, although matrix analogs exist.

The second advantage is more striking, and has to do basically with the numerics involved. The integration of the r and s equations is easier than is that of Eq. (8.20) simply because there are far fewer function evaluations involved. It is true, of course, that the imbedding formulation requires the switching back and forth between the equation for r and the equation for s but that is a minor programming problem. Moreover, both methods depend on finding where a particular function assumes a specified value. Now, from (8.23) we easily compute

$$r'(z) = [\sec^2\theta(z)]\theta'(z), \tag{8.24}$$

so that at the end point $z = x$ we have

$$|r'(x)| \geq |\theta'(x)|. \tag{8.25}$$

Thus the slope of the r curve is always at least as great (in absolute value) as that of the θ curve. Obviously, the same statement may be made about the function s. It is geometrically clear that the larger the slope of r or s at x, the greater is the accuracy of the calculation. (Indeed, examination of some of the work of [4] shows that the function θ often has a tendency to be quite flat, making the determination of the eigenlengths especially difficult using the Prüfer method.)

7. SOME NUMERICAL EXAMPLES

It is evident from our discussion that the method outlined can be used either for the calculation of eigenvalues or eigenlengths. In transport problems, for instance, the eigenlength is often the quantity of greatest interest. We shall present here three examples, the first having to do with eigenlengths, the others with eigenvalues. They originally appeared in [1].

Example 1

We consider the case of transport in a uniform rod geometry. Each end of the rod adjoints a reflecting material which sends some of the particles which emerge from the rod back into the system. Assuming binary fission ($f=b=1$), the corresponding equations for this model may be written

$$y''(z)+\sigma^2 y(z)=0, \tag{8.26a}$$

$$\beta_1 y'(0)-\sigma y(0)=0, \tag{8.26b}$$

$$y'(x)-\sigma\beta_2 y(x)=0. \tag{8.26c}$$

Here, β_1 and β_2 are the so-called albedo numbers for the reflecting materials, and σ is the usual macroscopic cross section. Hence, x is the critical length of the rod when (8.26) has a nonnegative solution. (See Problem 6 for more physical details.) This model has the distinct advantage of having an analytic solution so that a check on the numerics is possible.

In this, as well as in all other problems to be discussed in this section, calculations were done using a fourth-order Runge–Kutta scheme on a CDC 3600 computer. For the present case we chose $\beta_1=0.2$, $\beta_2=0.2$, and $\sigma=10.0$. We computed the eigenlength (physically, the critical size) using our method and an integration step of 10^{-2}. Relative error in the critical length, as compared with the analytic result, was about 6.7×10^{-4}. Reducing the step size to 5×10^{-3} improved this to 2.3×10^{-4}. These results are compatible with the accuracy of the integration routines used.

Example 2

As another example let us consider the nonlinear eigenvalue problem

$$y''(z)+(\lambda+\lambda^2 z^2)y(z)=0, \tag{8.27a}$$

$$y(-1)=y(1)=0. \tag{8.27b}$$

This may be considered as a perturbation of the much more standard harmonic oscillator problem in which the term $\lambda^2 z^2$ does not occur and whose solution is completely trivial. Actually, (8.27) has been studied by Collatz in [5], where he obtained bounds on λ_1. We obtained (see also Figure (8.2))

$$\lambda_1 = 1.9517,$$

$$\lambda_2 = 4.2861,$$

$$\lambda_3 = 7.5459.$$

The value for λ_1 falls well within the bounds $1.811 \leqslant \lambda_1 \leqslant 1.965$ provided by Collatz.

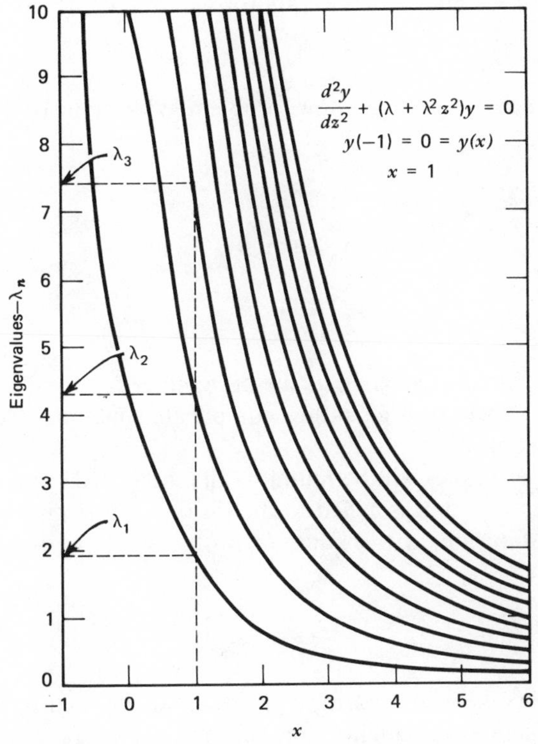

Figure 8.2.

Example 3

Our final example, strictly speaking, does not fit our theory, since the coefficient k in the differential operator has a zero. There seems to be considerable numerical evidence, and some strong theoretical indication, that our method is more widely applicable than the conditions we have placed upon the functions would have us believe. The equation

$$\frac{d}{dz}\left\{(1-z^7)\frac{dy(z)}{dz}\right\}+\lambda z^7 y(z)=0, \tag{8.28a}$$

$$y(0)=0, \qquad y(1) \text{ finite}, \tag{8.28b}$$

originally arose in some physical studies by Latzko [6]. It has since been investigated by a large number of researchers. (See the bibliography of [1] for a fairly extensive list.)

It is clear that to apply our method further information is needed concerning the behavior of the function $y(z)$ at $z=1$. A series expansion about the point $z=1$ shows that an appropriate boundary condition is

$$y'(1)=0. \tag{8.28c}$$

Using our method on the resulting problem we obtained the following values:

$$\lambda_1=8.728,$$

$$\lambda_2=152.45,$$

$$\lambda_3=435.2.$$

These are approximately as good as the best values found to date by other workers, many of whom used rather complicated methods devised especially for this problem.

Several other examples will be found by the interested reader in [1]. (See also [8].) Special attention is called to the third example of that paper and its possible relationship to the kinds of problems to be discussed in the next chapter.

8. SUMMARY

In this chapter we have turned to applications of invariant imbedding to eigenvalue problems. One might argue that the use of the term "invariant imbedding" is inappropriate here and that really all we have done is to make a rather obvious use of the very classical Riccati transformation. But it must also be noted that this particular employment of the Riccati substitution is apparently quite new, and would perhaps not have been thought of had the imbedding method not have been under intensive investigation. All this is even the more remarkable, since the Prüfer transformation device has been used for some time, and it is actually more complicated and, at least in many cases, less efficient than the algorithm we have developed.

Attention should also be called to the method referred to in the Russian literature as the "sweep" method or the "chasing" method [7]. It has many points of contact with our algorithm, but for some reason its application has largely been confined to the study of inhomogeneous second-order boundary value problems. It has not been used in the study of eigenvalue problems to the best of the authors' knowledge.

It should be remarked that the research necessary to extend the method we have been discussing to higher-order equations has not as yet been carried out in detail. As noted in Section 5 we *do* know how to apply invariant imbedding to higher-order equations, but various facets of its use in studying eigenvalue problems for such equations are not completely understood. Although the device presented in [8] seems eminently successful, there are quite a few nonpathological cases where it fails completely. The problem remains an open one.

PROBLEMS

1. Discuss the Fredholm alternative both from the viewpoint of differential equations and from the viewpoint of other operator equations. Attempt to understand the connection of the alternative to the concept of criticality in both one- and many-state systems.

2. Obtain derivations of the equation for r and the equation for s [Eqs. (8.3) and (8.6)].

3. Discuss in detail the fact that incidental to our analysis of the problem of Section 3 of this chapter we have also obtained an algorithm for solving the problem in which u is assigned to be zero at $z=0$ and $z=\tilde{z}_j$ [see Eq. (8.1d′)]. Observe that in Problems 12 and 13 of Chapter 3 it was noted that the method of invariant imbedding can be used even when the value of u is assigned at both ends of the interval under discussion. Relate these ideas. Does this suggest still other methods of computing eigenlengths and eigenvalues?

4. To what extent may the constants a_1, a_2, b_1, and b_2 in Eq. (8.8) not be arbitrary? Consider the problem defined by (8.8) and the transformation (8.11). Work out in detail the transformed problem and try to determine "good" choices of the a_{ij}. Carry through this type of analysis for (8.12) and suggest substitutions other than (8.13) that might be useful.

5. Try to develop the general ideas of this chapter for systems of linear equations. What aspects of our analysis carry over directly and what aspects seem less obvious? Give an example of total failure of the straightforward generalization.

6. Consider the transport Eqs. (1.8) and (1.9). Suppose the "rod" at $z=0$ adjoins a material which has the effect of reflecting back into the rod an expected total of β particles for each particle emergent from the rod at $z=0$. The number β is called the *albedo* at $z=0$. Show that this yields the condition $u(0)=\beta v(0)$. Place a reflecting material at $z=x$ and obtain a similar condition. Derive Eqs. (8.26) for the model under consideration in Example 1.

Generalize the albedo-type boundary conditions to the many state case. (See also Problem 14, Chapter 3.)

7. Physically one would expect that the "better" the reflectors on each end of the rod in the previous problem the shorter should be the critical length. Do some computations on Example 1 with various values of β_1 and β_2 to confirm this conjecture. Can you give an analytical proof in this case? Also study more general cases (f different from b; f and b nonconstant; the many-state case).

8. Example 3 has been stated in the way given because that is the manner in which the question arises physically. Confirm that the condition $y(1)$ finite does indeed lead to $y'(1)=0$.

REFERENCES

1. M. R. Scott, L. F. Shampine, and G. M. Wing, "Invariant Imbedding and the Calculation of Eigenvalues for Sturm–Liouville Systems," *Computing* **4**, 1969, 10–23.
2. F. Riesz and B. Sz.-Nagy, *Functional Analysis*, Frederick Ungar, New York, 1955.
3. E. L. Ince, *Ordinary Differential Equations*, Dover, New York, 1944.
4. P. B. Bailey, "Sturm–Liouville Eigenvalues via a Phase Function," *SIAM J. Appl. Math.* **14**, 1966, 242–249.
5. L. Collatz, "Monotonicity and Related Methods in Non-Linear Differential Equations," in *Numerical Solution of Nonlinear Differential Equations* (D. Greenspan, Ed.), Wiley, New York, 1966.
6. H. Latzko, "Wärmeübergang an Einem Turbulenten Flüssigkeitsoder Gasstrom," *Z. Angew. Math. Mech.* **1**, 1921, 268–290.
7. S. K. Gudonov and V. S. Ryabenki, *Theory of Differential Schemes*, Wiley, New York, 1964.
8. M. R. Scott, *Invariant Imbedding and its Applications to Ordinary Differential Equations*, Addison-Wesley, Reading, Mass., 1973.

9

SCHRÖDINGER-LIKE EQUATIONS

1. INTRODUCTION

In this chapter we shall apply the method of invariant imbedding to two types of problems that occur often in quantum mechanical studies. The first general class is frequently referred to as the "phase shift" problem, while the second is a particular form of the eigenvalue problem. In this latter case the method of the previous chapter is, within certain technical limits, applicable. However, the way in which we shall now make use of the imbedding scheme gives rise to a completely different characterization of the eigenvalues than the one we have been discussing. It seems particularly desirable to include a presentation of this method because the results illustrate so strikingly the way in which different kinds of imbeddings can yield completely different formulas. Although the phase shift problem is of less interest in itself, since our results can be derived in a rather more elementary way, we include it since once again, it demonstrates the versality of the method (see [1,2]).

While we have referred to quantum mechanics and the Schrödinger equation, we intend to avoid any genuine involvement with the physics from which the problems to be investigated originally arise. The reader need have no acquaintance with the subject of quantum mechanics to understand this chapter. Indeed, many of the questions we shall study arise in other contexts, and are of quite general interest.

Of the two basic questions to be investigated, the phase shift problem is by far the easier. Therefore, we shall treat it rather sketchily, leaving much to the reader. On the other hand, the eigenvalue results are rather difficult. For that reason, and because the authors are unaware of any alternative methods of obtaining the formulas, we shall include more of the details.

2. FORMULATION OF THE PHASE SHIFT PROBLEM

The equation we wish to study is

$$\frac{d^2u}{dt^2}+u=f(t)u, \qquad t\geqslant 0; \tag{9.1a}$$

$$u(0)=d_1, \tag{9.1b}$$

$$u'(0)=d_2, \tag{9.1c}$$

where

$$\int_0^{\infty}|f(t)|\,dt<\infty, \tag{9.2}$$

and the function $f(t)$ is assumed to be continuous and bounded on the entire interval $t\geqslant 0$. (This is actually a stronger condition than is required. See Problem 1.)

It is a classical result that Eq. (9.1a) has two fundamental solutions, which for large t behave like

$$u_1(t)=\sin t+o(1), \tag{9.3a}$$

$$u_2(t)=\cos t+o(1), \tag{9.3b}$$

so that the complete solution can be written in the form

$$\begin{aligned} u(t)&=c_1\sin t+c_2\cos t+o(1)\\ &=a\cos(t-\tilde{\theta})+o(1). \end{aligned} \tag{9.4}$$

Here, the quantity a is usually referred to as the amplitude and $\tilde{\theta}$ is called the phase shift. Obviously, these values are dependent on the initial conditions (9.1b) and (9.1c). Equally obviously, one way to determine a and $\tilde{\theta}$ is to integrate the system (9.1) out far enough in t until the computed solution has assumed essentially the asymptotic form (9.4). We seek a method of obtaining these quantities more directly, since there are difficulties involved with this procedure.

Our basic device is to imbed the given problem in a class of related problems, and then to obtain relationships between the quantity $\tilde{\theta}$ for one member of the class and the corresponding $\tilde{\theta}$ for an "adjacent" member; a similar idea can be devised to obtain information regarding a. Clearly, the basic philosophy of invariant imbedding is here, but the "class of related problems" must be chosen with considerable care and ingenuity.

We replace the initial conditions (9.1b, c) with more convenient ones:

$$u(x)=\cos\theta \tag{9.1b$'$}$$

and

$$u'(x)=\sin\theta. \tag{9.1c$'$}$$

Here $0 \leqslant x < \infty$, and θ is arbitrary. The fact that $u^2(x)+u'(x)^2=1$ does not result in any loss of generality because of the linearity of the original problem (9.1). It is useful to introduce the notation

$$u=u(t;x,\theta). \tag{9.5}$$

A prime will always denote differentiation of u with respect to the first argument. Furthermore, we choose to write the asymptotic form as

$$u(t;x,\theta)=a(x,\theta)\cos(t-x-\theta-\psi(x,\theta))+o(1). \tag{9.6}$$

Although (9.6) may appear initially a little awkward, its form has been carefully chosen to make many of our calculations easier algebraically. For ease of expression we shall sometimes refer to ψ as the "phase shift," despite the fact that this terminology does not agree with the classical one, nor with the term we have been using in our treatment up to this point.

It is now convenient to replace the problem (9.1a, b$'$, c$'$) by the equivalent integral equation

$$u(t;x,\theta)=\cos(t-x-\theta)+\int_x^t \sin(t-\omega)f(\omega)u(\omega;x,\theta)\,d\omega, \tag{9.7}$$

and to remark that classical differential equation theory assures us of the existence and the continuity of $u_x(t;x,\theta)$ and $u_\theta(t;x,\theta)$ in t, x, and θ. Finally, we shall often find it helpful to consider $u(t;x,\theta)$ to be defined only for $t \geqslant x$, despite the fact that the solution to (9.7) is known to exist for all $t \geqslant 0$.

3. A REPRESENTATION OF THE SOLUTION FOR LARGE t

In this section we shall obtain a more precise representation for the asymptotic solution of our problem. Our first result follows from a rather straightforward application of the Gronwall–Bellman inequality to (9.7). Details are left as a problem.

LEMMA 9.1. The function $u(t;x,\theta)$ satisfies the inequality

$$|u(t;x,\theta)| \leqslant \exp\left(\int_x^t |f(\omega)|\,d\omega\right) \leqslant m, \tag{9.8}$$

where m is a constant dependent only upon f.

Next, some manipulation of (9.7) yields

$$u(t;x,\theta)=\sin t\left\{\sin(x+\theta)+\int_x^\infty \cos\omega f(\omega)u(\omega;x,\theta)\,d\omega\right\}$$
$$+\cos t\left\{\cos(x+\theta)-\int_x^\infty \sin\omega f(\omega)u(\omega;x,\theta)\,d\omega\right\}$$
$$-\int_t^\infty \sin(t-\omega)f(\omega)u(\omega;x,\theta)\,d\omega=b(x,\theta)\sin t+c(x,\theta)\cos t$$
$$+\zeta(t;x,\theta). \tag{9.9}$$

From Eq. (9.9) one may obtain the following (see Problem 5).

LEMMA 9.2. The functions $b(x,\theta)$ and $c(x,\theta)$ have continuous partial derivatives with respect to both variables. Furthermore,

$$\lim_{t\to\infty}\zeta(t;x,\theta)=0 \tag{9.10}$$

uniformly in x and θ.

We can now state a fundamental theorem.

THEOREM 9.1. The solution $u(t;x,\theta)$ may be written in the form

$$u(t;x,\theta)=a(x,\theta)\cos(t-x-\theta-\psi(x,\theta))+\zeta(t;x,\theta), \tag{9.11}$$

where $a(x,\theta)$ and $\psi(x,\theta)$ have continuous partial derivatives with respect to both x and θ, and where

$$\lim_{x\to\infty}a(x,\theta)=1,$$
$$\lim_{x\to\infty}\psi(x,\theta)=0. \tag{9.12}$$

SKETCH OF PROOF. The proof of Theorem 9.1 is relatively easy in the light of the two lemmas we have given. We shall mention only a few troublesome points and leave the remaining work to the reader (see Problems 6 and 7). First, it is necessary to rewrite (9.9) in a form resembling (9.11).

$$u(t;x,\theta)=\sqrt{b^2(x,\theta)+c^2(x,\theta)}\left\{\frac{b(x,\theta)}{\sqrt{b^2(x,\theta)+c^2(x,\theta)}}\sin t + \frac{c(x,\theta)}{\sqrt{b^2(x,\theta)+c^2(x,\theta)}}\cos t\right\}+\zeta(t;x,\theta)$$

$$=a(x,\theta)\cos(t-x-\theta-\psi(x,\theta))+\zeta(t;x,\theta). \tag{9.13}$$

Clearly, this will not be possible if b and c vanish simultaneously. It is not hard to rule out this possibility. Moreover, we note in passing that $a(x,\theta)$ can always be chosen to be positive. The continuous differentiability of a now follows from that of b and c. To see that $a(x,\theta)\to 1$ as $x\to\infty$ one must go back to (9.9) to find the specific form of b and c.

Finally, recalling that we always assume $t\geqslant x$, we can see from (9.9) that

$$u(t;x,\theta)=\cos(t-x-\theta)+\hat{\zeta}(t;x,\theta), \tag{9.14}$$

where $\hat{\zeta}(t;x,\theta)\to 0$ as $x\to\infty$. Thus we may require that $\psi(x,\theta)\to 0$ as $x\to\infty$. It is now clear that ψ may be defined in such a way that it, too, is continuously differentiable in both its variables.

4. PARTIAL DIFFERENTIAL EQUATIONS FOR a AND ψ

The properties of a and ψ developed in the preceding section may now be exploited, and the promised imbedding accomplished. Choose $\Delta>0$ and consider $u(x+\Delta;x,\theta)$. From Eq. (9.7) we readily compute

$$u(x+\Delta;x,\theta)=\cos\theta+\Delta\sin\theta+o(\Delta). \tag{9.15a}$$

Similarly

$$u'(x+\Delta;x,\theta)=\sin\theta+\Delta(f(x)-1)\cos\theta+o(\Delta). \tag{9.15b}$$

Now we wish to consider a problem like (9.1a, b′, c′) but on the interval $x+\Delta\leqslant t<\infty$, using (9.15a, b) as initial values. Unfortunately, this cannot quite be done, because there is no assurance that

$$u^2(x+\Delta;x,\theta)+[u'(x+\Delta;x,\theta)]^2=1; \tag{9.16}$$

in fact, this is very probably *not* the case.

To carry out the imbedding we need the "adjacent problem" to be of the same form as the original. To overcome the difficulty, we define a quantity $\eta(x,\theta)$ as follows:

$$\begin{aligned} u(x+\Delta;x,\theta)&=\eta(x,\theta)u(x+\Delta;x+\Delta,\theta')=\eta(x,\theta)\cos\theta', \\ u'(x+\Delta;x,\theta)&=\eta(x,\theta)u'(x+\Delta;x+\Delta,\theta')=\eta(x,\theta)\sin\theta'. \end{aligned} \tag{9.17}$$

Now a relatively simple calculation using (9.15a, b) yields

$$\theta'=\theta+(f(x)\cos^2\theta-1)\Delta+o(\Delta), \tag{9.18a}$$

$$\eta=1+\cos\theta\sin\theta f(x)\Delta+o(\Delta). \tag{9.18b}$$

Finally, we make use of the linearity of our original initial value problem to obtain for all $t\geqslant x+\Delta$,

$$u(t;x,\theta)\equiv\eta(x,\theta)u(t,x+\Delta,\theta'), \tag{9.19}$$

so that

$$\begin{aligned} &a(x,\theta)\cos(t-x-\theta-\psi(x,\theta))+\zeta(t;x,\theta) \\ &\quad=\eta(x,\theta)\{a(x+\Delta,\theta')\cos(t-x-\Delta-\theta'-\psi(x+\Delta,\theta'))\}+\zeta(t;x+\Delta,\theta'). \end{aligned} \tag{9.20}$$

Hence,

$$-\theta-\psi(x,\theta)=-\Delta-\theta'-\psi(x+\Delta,\theta') \tag{9.21a}$$

$$a(x,\theta)=\eta(x,\theta)a(x+\Delta,\theta'). \tag{9.21b}$$

In the preceding section we showed that the various partial derivatives of ψ and a exist and are continuous. Hence Eqs. (9.21) lead to

$$\frac{\partial\psi}{\partial x}+\frac{\partial\psi}{\partial\theta}(f(x)\cos^2\theta-1)=-f(x)\cos^2\theta, \tag{9.22a}$$

$$\frac{\partial a}{\partial x}+\frac{\partial a}{\partial\theta}(f(x)\cos^2\theta-1)=\cos\theta\sin\theta f(x)a(x,\theta). \tag{9.22b}$$

It is perhaps appropriate to stop for just a moment and ask whether we have truly accomplished anything. The system (9.1) is an initial value problem, but it involves only one *ordinary* linear differential equation. We have now obtained two linear *partial* differential equations—suggesting perhaps an overall loss of ground, despite several pages of work. The new equations, however, are for precisely the quantities we are seeking. In the next section we shall show how they may be used to good advantage.

5. SOLUTION OF THE PARTIAL DIFFERENTIAL EQUATIONS FOR a AND ψ

The system (9.22) may be easily handled by the method of characteristics. From (9.22a) we get

$$\frac{dx}{ds}=1, \qquad \frac{d\theta}{ds}=f(x)\cos^2\theta-1, \qquad \frac{d\psi}{ds}=-f(x)\cos^2\theta, \tag{9.23}$$

so that, choosing $s=x$, and $\theta=\theta_0$ when $s_0=x_0$, we have

$$\psi(x_0,\theta_0)=\int_{x_0}^{\infty} f(s)\cos^2\theta(s)\,ds. \tag{9.24}$$

In the last equation $\theta(s)$ is obtained by integrating

$$\frac{d\theta}{ds}(s)=f(s)\cos^2\theta(s)-1 \tag{9.25}$$

with $\theta=\theta_0$ at $s=s_0$. (In most cases the integration must, of course, be done numerically.) The condition (9.2) insures the convergence of the integral in (9.24).

Obviously, a similar analysis may be made of (9.22b). The following theorem summarizes our result.

THEOREM 9.2. Let $u(t;x,\theta)$ be the solution to the problem (9.1a, b′, c′). Then

$$u(t;x,\theta)=a(x,\theta)\cos(t-x-\theta-\psi(x,\theta))+\zeta(t;x,\theta),$$

where

$$\lim_{t\to\infty}\zeta(t,x,\theta)=0$$

the limit being uniform in x and θ. Moreover,

$$\psi(x,\theta)=\int_x^{\infty} f(x)\cos^2\hat{\theta}\,(s)\,ds,$$

$$a(x,\theta)=\exp\left\{\tfrac{1}{2}\int_x^{\infty} f(x)\sin 2\hat{\theta}\,(s)\,ds\right\},$$

and $\hat{\theta}(s)$ is given as the solution to the nonlinear initial value problem

$$\frac{d\hat{\theta}}{ds}(s)=f(s)\cos^2\hat{\theta}\,(s)-1,$$

$$\hat{\theta}\,(x)=\theta.$$

6. REMARKS ON THE PHASE SHIFT PROBLEM

As indicated in the first section of this chapter, the result we have just obtained by the invariant imbedding method can be derived in other ways. We shall sketch one (for details, see [3]). Since most authors are interested in the quantum mechanical problem, it is customary in the literature to take $x=0$ and $\theta=\pi/2$. One then may try to write the solution for the resulting system (9.1) in the form

$$u(t)=\tilde{a}(t)\sin(t+\tilde{\eta}(t)), \qquad u'(t)=\tilde{a}(t)\cos(t+\tilde{\eta}(t)). \tag{9.26}$$

If (9.26) is inserted in the differential equation satisfied by $u(t)$ it is discovered that $\tilde{\eta}(t)$ must satisfy the equation

$$\frac{d\tilde{\eta}}{dt}=-f(t)\sin^2(t+\tilde{\eta}(t)), \tag{9.27a}$$

$$\tilde{\eta}(0)=0. \tag{9.27b}$$

A careful but relatively elementary analysis eventually reveals that

$$\lim_{t\to\infty}\tilde{\eta}(t)=-\psi\left(0,\frac{\pi}{2}\right). \tag{9.28}$$

(For the details the reader is referred to [1] and to the problems.) Thus the phase shift problem is resolved in a fairly elementary way. However, a similar device does not seem to be available for the eigenvalue problem that we shall treat shortly. That question can be handled by exactly the same sort of imbedding as we have just used in the phase shift case, although the technical details are considerably more difficult.

One immediately wonders what can be done with operators similar to, but more complicated than, the one in (9.1a). First efforts in this direction were made in [4], where the imbedding method was applied to a class of differential equations whose left-hand sides are different from that of (9.1a), but which are still somewhat specialized to the needs of the quantum mechanician. Some interesting problems were encountered. A much more general treatment appears in [5]. There the differential operator is replaced by

$$\frac{d^2}{dt^2}u+(1+g(t))u(t)=f(t)u(t), \tag{9.29}$$

where g is absolutely continuous, g^2 is integrable at infinity, $|g'|$ is also integrable at infinity, and $g\to 0$ at $t\to\infty$. The function f satisfies the conditions of this chapter. The assumptions on g allow one to write down an asymptotic expression for the fundamental solutions to the homogeneous problem, and then the method works *formally* like the one we have outlined. There are, however, many rather unpleasant questions of convergence of infinite integrals, and the like, which very considerably complicate the analysis. Indeed, it is precisely because many matters with respect to the material of this chapter are technically quite difficult, as well as not being analytically obvious, that we have chosen a more rigorous approach than we have used in many of the previous parts of this book.

Some experimental calculations have been done on the phase shift problem, often with rather disappointing results. Integrals frequently tend to converge rather slowly, integration intervals must often be taken quite small, and so on. Nevertheless, the method is probably at least as good as the straightforward approach mentioned in the early part of Section 2, and very likely superior in some instances. As far as we know, no thorough study has been made of the relative efficacies of the methods.

7. FORMULATION OF THE EIGENVALUE PROBLEM

We now turn our attention to the system

$$\frac{d^2y}{dt^2}-\lambda^2y(t)=f(t)y(t), \qquad 0\leqslant t<\infty, \tag{9.30a}$$

$$y(0)=d_1, \tag{9.30b}$$

$$y'(0)=d_2, \tag{9.30c}$$

where f is as in Section 2 and λ is a real positive number. Again it is a classical result that there are two fundamental solutions to (9.30a) having the form

$$y_1(t)=e^{\lambda t}(1+o(1)), \tag{9.31a}$$

$$y_2(t)=e^{-\lambda t}(1+o(1)), \tag{9.31b}$$

as $t\to\infty$. Hence the solution to the initial value problem can be written

$$\begin{aligned} y(t)=a_1y_1(t)+a_2y_2(t)&=a_1e^{\lambda t}(1+o(1))\\ &+a_2e^{-\lambda t}(1+o(1)). \end{aligned} \tag{9.32}$$

In many physical problems one is interested in those values of λ (if any) such that $y(t)\to 0$ as $t\to\infty$. We note from Eq. (9.32) that this will be the case if a_1, which is actually a function of λ, is zero. Thus, instead of the usual technique employed in the solution of such problems, which requires the calculation of the eigenfunction y itself, we shall attempt to find the functional dependence of a_1 on λ and then find the (positive) zeros of this function a_1. In doing so, we shall lose most information about the eigenfunctions themselves, but they are often of only secondary interest.

The imbedding we choose is essentially that of the earlier sections on the phase shift problem: namely, we study the family of problems defined by (9.30a)

$$y(x)=\cos\theta, \qquad 0\leqslant x<\infty, \tag{9.30b$'$}$$

and

$$y'(x)=\sin\theta, \qquad 0\leqslant x<\infty. \tag{9.30c$'$}$$

We shall usually write

$$y=y(t;x,\theta), \tag{9.33}$$

and conform in general to the agreements made concerning the earlier u notation. (It might be noted that y depends also on λ, but we choose not to exhibit that explicitly.)

The integral equation analogous to (9.7) is

$$\begin{aligned} y(t;x,\theta)=&\cos\theta\cosh\lambda(t-x)+\lambda^{-1}\sin\theta\sinh\lambda(t-x)\\ &+\lambda^{-1}\int_x^t \sinh\lambda(t-\omega)f(\omega)y(\omega;x,\theta)\,d\omega. \end{aligned} \tag{9.34}$$

In examining this equation it is convenient to introduce the new function

$$z(t;x,\theta)=e^{-\lambda(t-x)}y(t;x,\theta) \tag{9.35}$$

so that (9.34) becomes

$$\begin{aligned} z(t;x,\theta)=&\frac{1}{2}\left(\cos\theta+\frac{\sin\theta}{\lambda}\right)+\frac{1}{2}\left(\cos\theta-\frac{\sin\theta}{\lambda}\right)e^{-2\lambda(t-x)}\\ &+\frac{1}{2\lambda}\int_x^t[1-e^{-2\lambda(t-\omega)}]f(\omega)z(\omega;x,\theta)\,d\omega. \end{aligned} \tag{9.36}$$

This allows an easy estimate of z:

$$|z(t;x,\theta)|\leqslant\left(1+\frac{1}{\lambda}\right)+\frac{1}{2\lambda}\int_x^t|f(\omega)||z(\omega;x,\theta)|\,d\omega, \tag{9.37}$$

and the Gronwall–Bellman inequality yields (see Problems 3 and 4) the following lemma.

LEMMA 9.3. The function $z(t;x,\theta)$ satisfies the inequality

$$|z(t;x,\theta)| \leqslant \left(1+\frac{1}{\lambda}\right)\exp\left\{\frac{1}{2\lambda}\int_x^t |f(\omega)|\,d\omega\right\} \leqslant m', \tag{9.38}$$

where m' is a constant independent of x and θ and dependent only upon f and λ.

Next we state a lemma that is analogous to Lemma 9.2. The proof parallels that of the previous result except for one rather delicate portion. We leave the whole matter as a problem.

LEMMA 9.4. The function z may be expressed in the form

$$z(t;x,\theta)=\tilde{a}(x,\theta)+\tilde{\xi}(t;x,\theta), \tag{9.39}$$

where $\tilde{a}$ has a continuous first partial derivatives with respect to both x and θ. Moreover,

$$\tilde{a}(x,\theta)=\tfrac{1}{2}\left(\cos\theta+\frac{\sin\theta}{\lambda}\right)+\frac{1}{2\lambda}\int_x^\infty f(\omega)z(\omega;x,\theta)\,d\omega, \tag{9.40a}$$

$$\lim_{t\to\infty}\tilde{\xi}(t;x,\theta)=0, \tag{9.40b}$$

and the limit is uniform with respect to x and θ.

It is now simple to establish the basic theorem of this section.

THEOREM 9.3. The solution $y(t;x,\theta)$ of (9.30a, b′, c′) may be written in the form

$$y(t;x,\theta)=e^{\lambda(t-x)}\left\{\tilde{a}(x,\theta)+\tilde{\xi}(t;x,\theta)\right\}, \tag{9.41}$$

where $\tilde{a}$ and $\tilde{\xi}$ are as in Lemma 9.4. Furthermore,

$$\lim_{x\to\infty}\tilde{a}(x,\theta)=\tfrac{1}{2}\left(\cos\theta+\frac{\sin\theta}{\lambda}\right) \tag{9.42}$$

and the limit is uniform with respect to θ.

We shall find in the analysis that follows that (9.42) is not a convenient condition with which to work. At times we shall utilize the function

$$\tilde{b}(x,\theta)=\tilde{a}(x,\theta)-\tfrac{1}{2}\left(\cos\theta+\frac{\sin\theta}{\lambda}\right), \tag{9.43}$$

so that

$$y(t;x,\theta)=e^{\lambda(t-x)}\left[\left\{\tilde{b}(x,\theta)+\tfrac{1}{2}\left(\cos\theta+\frac{\sin\theta}{\lambda}\right)\right\}+\tilde{\zeta}(t;x,\theta)\right] \tag{9.44}$$

and

$$\lim_{x\to\infty}\tilde{b}(x,\theta)=0. \tag{9.45}$$

(It should be noted that quantities such as $\tilde{a}$, $\tilde{b}$, and $\tilde{\zeta}$ all depend on λ as well as on the more customary variables. We have chosen to suppress this dependence up to the present in order to emphasize the correspondence with the various functions in the phase shift problem. Later, when we begin study of the eigenvalue problem per se, we shall have to modify our notation to take the λ parameter into account.)

8. A PARTIAL DIFFERENTIAL EQUATION FOR $\tilde{b}$ AND ITS "SOLUTION"

We next seek a partial differential equation for $\tilde{b}$ analogous to the one obtained for $a(x,\theta)$ in the phase shift problem. Here the analysis is parallel to that done previously and we leave the details completely to the reader (see Problem 13). It actually seems somewhat easier to derive an equation for $\tilde{a}$ directly and then obtain the desired result for $\tilde{b}$ by using the definition (9.43). The result is

$$\frac{\partial\tilde{b}}{\partial x}+[(f(x)+1+\lambda^2)\cos^2\theta-1]\frac{\partial\tilde{b}}{\partial\theta}$$
$$=[\lambda-(f(x)+1+\lambda^2)\cos\theta\sin\theta]\tilde{b}(x,\theta)-\frac{f(x)}{2\lambda}\cos\theta. \tag{9.46}$$

Again applying the method of characteristics with

$$\frac{dx}{ds}=1,\qquad\frac{d\theta}{ds}=(f(x)+1+\lambda^2)\cos^2\theta(x)-1,$$
$$\frac{d\tilde{b}}{ds}=\left[\lambda-(f(x)+1+\lambda^2)\cos\theta(x)\sin\theta(x)\tilde{b}-\frac{f(x)}{2\lambda}\cos\theta(x)\right], \tag{9.47}$$

we obtain *formally*

$$\tilde{b}(x,\theta)=\frac{1}{2\lambda}\int_x^{\infty}\left[f(s)\cos\tilde{\theta}(s)\right]$$

$$\cdot\exp\left\{\int_x^s\left[\left(f(p)+1+\lambda^2\right)\cos\tilde{\theta}(p)\sin\tilde{\theta}(p)-\lambda\right]dp\right\}ds, \quad (9.48a)$$

$$\frac{d\tilde{\theta}}{ds}(s)=\left(f(s)+1+\lambda^2\right)\cos^2\tilde{\theta}(s)-1, \quad (9.48b)$$

$$\tilde{\theta}(x)=\theta. \quad (9.48c)$$

Unfortunately, the infinite integral that appears in (9.48a) is at the moment only formal; its convergence is not the least bit obvious. The difficult arises from the exponential term. That part of the integrand which contains the function $f(p)$ causes no problem because of the assumptions concerning f. The remaining expression

$$\int_s^x\left[\left(1+\lambda^2\right)\cos\tilde{\theta}(p)\sin\tilde{\theta}(p)-\lambda\right]dp \quad (9.49)$$

is troublesome. If $\lambda=1$ then (9.49) reduces to

$$\int_s^x\left(\sin 2\tilde{\theta}(p)-1\right)dp, \quad (9.50)$$

which is nonpositive. In this case, then, there is no question concerning the convergence of the expression for $\tilde{b}$. However, for any other value of λ it is not at all apparent—and perhaps it is not even true—that (9.49) is nonpositive. It has been possible thus far to overcome this problem only by rather devious means, which we shall now outline.

We return to the problem (9.30a, b′, c′) and make the following transformations:

$$\lambda t=z, \qquad y(t)=y(z/\lambda)=w(z). \quad (9.51)$$

Our basic problem then becomes

$$\frac{d^2w}{dz^2}(z)-w(z)=\frac{1}{\lambda^2}f(z/\lambda)w(z), \quad (9.52a)$$

$$w(\lambda x)=\cos\theta. \quad (9.52b)$$

$$\frac{d}{dz}w(z)\big|_{z=\lambda x}=\frac{\sin\theta}{\lambda}. \quad (9.52c)$$

Now define

$$\tilde{\theta} = \tan^{-1}\left(\frac{1}{\lambda}\tan\theta\right) \tag{9.53a}$$

$$\xi = \frac{\sqrt{\lambda^2\cos^2\theta + \sin^2\theta}}{\lambda}. \tag{9.53b}$$

Equations (9.52b, c) may be rewritten

$$w(\lambda x) = \xi\cos\tilde{\theta} \tag{9.54a}$$

and

$$\frac{d}{dz}w(z)|_{z=\lambda x} = \xi\sin\tilde{\theta}. \tag{9.54b}$$

We recall that only the eigenvalues of the original problem are really of interest to us, and that all problems being investigated are linear. Hence, it suffices to examine the new system

$$\frac{d^2w(z)}{dz^2} - w(z) = F(z)w(z), \tag{9.55a}$$

$$w(\tilde{x}) = \cos\tilde{\theta}, \tag{9.55b}$$

$$w'(\tilde{x}) = \sin\tilde{\theta}, \tag{9.55c}$$

where

$$\tilde{x} = \lambda x, \qquad F(z) = \frac{1}{\lambda^2}f\left(\frac{z}{\lambda}\right). \tag{9.56}$$

Using the methods of this section and the preceding one we discover, using an obvious notation,

$$\tilde{b}(\tilde{x},\tilde{\theta}) = \tfrac{1}{2}\int_{\tilde{x}}^{\infty}\left[F(s)\cos\hat{\theta}(s)\right] \cdot\exp\left\{-\int_{\tilde{x}}^{s}\left[1-\sin 2\hat{\theta}(p)\left(1+\frac{F(p)}{2}\right)\right]dp\right\}ds, \tag{9.57a}$$

$$\frac{d\hat{\theta}}{ds} = F(s)\cos^2\hat{\theta}(s) + \cos 2\hat{\theta}(s), \tag{9.57b}$$

$$\hat{\theta}(\tilde{x}) = \tilde{\theta}. \tag{9.57c}$$

Suprisingly enough, there is now no convergence problem; the integral in (9.57a) is very well behaved, and our difficulties seem to have been eliminated. There is a hidden disadvantage, however: namely, the eigenvalue parameter λ is now concealed in F, $\tilde{x}$, and $\tilde{\theta}$. Since ultimately we shall obviously have to resort to numerical methods, this feature may make the programming for machine calculation much more complex. We have resolved our theoretical difficulty, but we may have increased our practical one!

Before turning to an overall resolution of these problems, it is appropriate to discuss what we *hope* will be the case. The solution to the system (9.55) obviously has the form

$$w(z)=e^{z-\tilde{x}}\left\{\tilde{b}\,(\tilde{x},\tilde{\theta},1)+\frac{\cos\tilde{\theta}+\sin\tilde{\theta}}{2}+\tilde{\zeta}\,(z,\tilde{x},\tilde{\theta},1)\right\}. \tag{9.58}$$

(It is now convenient and desirable to make $\tilde{b}$, $\tilde{\zeta}$, and so on, possess an additional dependence, which we do by simply adding an extra argument. The last argument in these functions now indicates the coefficient of the undifferentiated term on the left side of the fundamental differential equation under study. In the case of the problem defined by (9.55) this argument is simply unity; for the equation (9.30) it is λ.) It is reasonable to conjecture that the eigenvalues λ of the original problem are determined by the condition

$$\tilde{b}\,(\tilde{x},\tilde{\theta},1)+\frac{\cos\tilde{\theta}+\sin\tilde{\theta}}{2}=0 \tag{9.59}$$

or, in the case that (9.48a) *is* well defined, by

$$\tilde{b}\,(x,\theta,\lambda)+\frac{\lambda\cos\theta+\sin\theta}{2\lambda}=0. \tag{9.60}$$

It is not entirely clear that this is the case, however, since it is at least conceivable that the terms $e^{z}\tilde{\zeta}$ or $e^{\lambda t}\tilde{\zeta}$ will make some contribution.

9. RESOLUTION OF THE DIFFICULTIES

In this section we shall overcome the two troublesome points that arose in Section 8. Let us begin with the one mentioned in the last sentence of that section. We may state our result as a lemma.

LEMMA 9.5. A necessary and sufficient condttion that λ be an eigenvalue of the problem (9.30) is that $\tilde{b}(\tilde{x},\tilde{\theta},1)+(\cos\tilde{\theta}+\sin\tilde{\theta})/2=0$. In the event that $\tilde{b}(x,\theta,\lambda)$ exists, this condition may be restated as $\tilde{b}(x,\theta,\lambda)+(\lambda\cos\theta+\sin\theta)/2\lambda=0$.

PROOF. We shall only consider the case for $\tilde{b}(\tilde{x},\tilde{\theta},1)$. Suppose that $\tilde{b}(\tilde{x},\tilde{\theta},1)+(\cos\tilde{\theta}+\sin\tilde{\theta})/2\neq 0$. From Eqs. (9.44) it follows at once that $y(z;\tilde{x},\tilde{\theta},1)$ becomes infinite as t goes to infinity. The necessity is hence evident.

The sufficiency is a little more delicate. The original proof (see [2]) was quite complicated and used rather deep results from the theory of eigenvalue problems on an infinite domain. The following much more transparent proof is given in [5]. Recall Eq. (9.32), and suppose that $\tilde{b}(\tilde{x},\tilde{\theta},1)+(\cos\theta+\sin\theta)/2=0$. It follows that

$$y(z;\tilde{x},\tilde{\theta},1)e^{-z}=\tilde{\zeta}\,(z;\tilde{x},\tilde{\theta},1)e^{-\tilde{x}}=o(1). \tag{9.61}$$

But then it can only be that $a_1=0$ and so $y(z;\tilde{x},\tilde{\theta},1)\to 0$ as $z\to\infty$. This completes the proof.

The remaining difficulty in the preceding section involves the convergence of the integral expression for $\tilde{b}$ for $\lambda\neq 1$. This may be overcome in two ways. First, we may add further hypotheses concerning the function f in order to insure the good behavior of the integral. This has the distinct disadvantage of further restricting the class of problems under study. Second, and this is the course we shall follow, we may bear in mind that our ultimate goal is the numerical calculation of the eigenvalues of the system (9.30), and recognize that in such computations certain approximations are an absolute and unavoidable necessity.

We define the function

$$\tilde{b}_T(\tilde{x},\tilde{\theta},1)=\tfrac{1}{2}\int_{\tilde{x}}^{T}F(s)\cos\hat{\theta}\,(s)\exp\{\cdots\}\,ds,\qquad T>\tilde{x}. \tag{9.62}$$

This is actually the function that we would compute in a practical case, and, of course, $\tilde{b}_T\to\tilde{b}$ as $T\to\infty$. Denote the λ-zeros of $\tilde{b}_T$ by $\lambda_j(T)$. It is known from classical theory that the eigenvalues of (9.30) are simple; write the first N of them: $\lambda_1>\lambda_2>\cdots>\lambda_N>0$. (See [6].) Now it is easily verified that for given N and given $\epsilon>0$, there exists $T(\epsilon,N)$ such that for $T\geqslant T(\epsilon,N)$

$$|\lambda_j-\lambda_j(T)|<\epsilon,\qquad j=1,2,\ldots,N. \tag{9.63}$$

The truncation of the integral in the definition of $\tilde{b}_T$ [see Eq. (9.62)] is precisely equivalent to saying that

$$F(s)=\frac{1}{\lambda^2}f\left(\frac{s}{\lambda}\right)=0 \tag{9.64}$$

for $s>T$, or that $f(s)=0$ for $s>T/\lambda$. Hence, if we assume $f(s)=0$ for $s>T'=T/\lambda_N$ then we obtain the approximation (9.63).

What does all this mean in terms of the original problem (9.30a, b′, c′)? It means that we have replaced that problem by

$$\frac{d^2y}{dt^2}-\lambda^2y=\hat{f}(t)y, \tag{9.65a}$$

$$y(x)=\cos\theta, \tag{9.65b}$$

$$y'(x)=\sin\theta, \tag{9.65c}$$

where

$$\begin{cases}\hat{f}(t) & =f(t), \quad t\leqslant T',\\ \hat{f}(t) & =0, \quad t>T'\end{cases} \tag{9.66}$$

(Although our analysis has been for continuous f, piecewise continuity suffices; see Problem 1). But for this system the integral expression for the $\tilde{b}$ function is simply

$$\tilde{b}_{T'}(x,\theta,\lambda)=\frac{1}{2\lambda}\int_x^{T'}f(s)\cos\hat{\theta}(s)\exp\{\cdots\}\,ds, \qquad T'>x, \tag{9.67}$$

and here there is obviously no question concerning convergence. Moreover, this is just the integral we would use in *computations* if we knew that (9.48a) *did* converge. We can summarize these ideas in a formal theorem.

THEOREM 9.4. The eigenvalues of the system (9.30) can be approximated by calculating the λ-zeros of either

$$\tilde{b}_{T'}(x,\theta,\lambda)+\frac{\lambda\cos\theta+\sin\theta}{2\lambda}, \tag{9.68a}$$

or

$$\tilde{b}_T(\tilde{x},\tilde{\theta},1)+\frac{\cos\tilde{\theta}+\sin\tilde{\theta}}{2}. \tag{9.68b}$$

It must be admitted that the method we have just employed seems rather devious and contrived. A direct proof of the convergence of the integral defining the original $\tilde{b}$ function is much to be desired.

10. SOME NUMERICAL EXAMPLES

It was mentioned at the end of Section 6 that the phase shift formulas derived earlier in this chapter proved somewhat disappointing numerically. Quite the opposite is true with respect to the results obtained from the eigenvalue equations we have just derived. We shall mention just two examples, referring the reader to [2] for further numerical information. Both of the sets of data to be presented were obtained on a rather early model computing machine, the IBM 709. It seems highly likely that the "new generation" computers would produce even more satisfactory numbers, and with much less expenditure of time.

Example 1

$$f(t)=\frac{\alpha(1-\alpha)}{4\cosh^2(t/2)}.$$

This particular choice of f is especially interesting since an analytic solution to the problem (9.30) is available [6]. The eigenvalues are given in the case $x=0$, and $\theta=\frac{1}{2}\pi$, by

$$\lambda=\frac{\alpha}{2}-1-r, \qquad r=0,1,\ldots,\left[\frac{\alpha}{2}-\frac{3}{2}\right]. \tag{9.69}$$

Two cases were calculated, $\alpha=5$ and $\alpha=11$.

Here, as in the next example to be discussed, $\tilde{b}$ was tabulated as a function of λ. A fourth order Runge–Kutta scheme was used to carry out the integration of the system (9.48). First, a rough plot of $\tilde{b}+1/2\lambda$ was made and then more accurate calculations were carried out when a zero had been isolated. All values of λ were computed correct to four decimal places with very little difficulty.

Example 2

$$f(t)=\begin{cases} -v_0, & 0\leqslant t\leqslant t_0,\\ 0, & t>t_0.\end{cases}$$

This rather obvious "square well potential" problem also can be solved exactly. The eigenvalues are given as the solutions to the transcendental equation (once again assuming $x=0$ and $\theta=\frac{1}{2}\pi$)

$$\tan\left[t_0\sqrt{v_0-\lambda^2}\,\right]=-\frac{\sqrt{v_0-\lambda^2}}{\lambda}. \tag{9.70}$$

Again two cases were tried. In the first case the choice $v_0=4$, $t_0=10$ was used. Since the expression $\tilde{b}+1/2\lambda$ turned out to be quite small here, the machine being used necessitated the use of double precision arithmetic in order to obtain reasonable answers. In the second instance, with $v_0=4$ and $t_0=5$, this difficulty disappeared and single precision calculation proved entirely satisfactory. Results are shown in the table below.

Case 1		Case 2	
Calculated	Exact	Calculated	Exact
1.9769	1.9775	1.9150	1.9169
1.9094	1.9085	1.6483	1.6462
1.7856	1.7881	1.0850	1.0780
1.6076	1.6054		
1.3391	1.3375		
.9234	.9203		

11. SOME REMARKS ON THE EIGENVALUE PROBLEM

Just as in the case of the phase shift problem, we may reasonably ask if the operator (9.30a) may not be generalized somewhat. This question was first studied in [5]. There the equation we have investigated was replaced by

$$\frac{d^2y}{dt^2}-(\lambda^2+g(t))y(t)=f(t)y(t), \tag{9.71}$$

where the function g has the properties described in Section 6. Results similar to those we have just derived were obtained, although the analysis was somewhat complicated. Moreover, numerical experiments were carried out with various f and g functions and the results were very satisfactory.

It is also interesting to note that our overall approach produces quite easily some already basically known information concerning the distribution of eigenvalues. Recall that $\tilde{b}(\tilde{x},\tilde{\theta},1)$ is a function of λ. A careful analysis reveals that it is actually an *analytic* function of λ provided $\mathrm{Re}(\lambda)>0$. From this observation it follows that the eigenvalues $\lambda>0$ are bounded, isolated, and simple. (This result holds for the general case described in the above paragraph as well as for the special operator we have been studying.) (See [6]; also Problem 15.)

12. SUMMARY

In this chapter we have concentrated on the application of the imbedding method to the resolution of certain problems in differential equation theory which are posed on the semi-infinite line. We have obtained results for both the so-called phase shift problem and for the classical eigenvalue problem. Especially in the latter case the invariant imbedding method seems to be a particularly potent weapon.

Various questions immediately come to mind, however. In our analysis we have always assumed the functions f and g are appropriately "small" at infinity. There are many interesting problems in which this is not the case. We have also dealt only with second order differential operators in which the first derivative term is missing. While a suitable transformation can always be used to achieve this situation, it is reasonable to ask if the method may be successfully applied when this derivative is present. We have considered only problems in which f and g are continuous [although it has been noted that piecewise continuity suffices (see Problem 1)]. Can the method be applied to cases in which these functions become infinite in the vicinity of the initial point, $z=x$? In [5] Hagin has examined some of these matters both analytically and numerically. The numerical results are frequently very striking, and one is tempted on the basis of them to make a crude guess that the worse the behavior of the problem when viewed from the classical viewpoint the better is its behavior when the imbedding method is applied. It should be emphasized that this remark is purely heuristic, and a good deal of further analysis and numerical experimentation is called for.

It is interesting to note that in [7] Hagin also looked into the possibility of determining something about the behavior of the eigenfunction itself in the neighborhood of infinity. In particular, he found results for the coefficient of the principal term of the eigenfunctions of the Hermite problem. Numerical results were somewhat disappointing, with answers obtainable only to two or three significant figures. (The corresponding eigenvalues were easily computed to five or six figures.) Surely more investigation in this direction is desirable.

To the best of the authors' knowledge, no study of the kind described in this chapter has been made for differential operators of order higher than two. This appears to be a quite difficult problem, but since such operators do regularly arise in practice, research on them could well be very rewarding.

PROBLEMS

1. Verify that the assumption that $f(t)$ is continuous is not needed for much of the work of the chapter. Show that f may indeed be piecewise continuous with a finite number of discontinuities. Can an infinite number of discontinuities be allowed provided they have no (finite) point of accumulation? Observe that the assumption of boundedness on f is not nearly so easily dropped. Where do the difficulties arise?

2. Obtain (9.7) and show that it and (9.1a, b′, c′) are indeed equivalent.

3. The Gronwall–Bellman inequality states the following: Let $w(z)$ and $q(z)$ be positive, piecewise continuous functions and let c be a positive constant. Suppose

$$w(z) \leqslant c + \int_0^z w(t)q(t)\,dt,$$

then

$$w(z) \leqslant c\exp\left\{\int_0^z q(t)\,dt\right\}.$$

Prove the inequality. Can the condition w positive be replaced by w merely nonnegative? Can the positivity conditions on q and c be relaxed similarly?

4. Using the result of Problem 3 establish (9.8) and (9.38).

5. Establish Lemma 9.2. Observe that the asymptotic forms (9.3) are obtainable from (9.9).

6. Verify that the quantities $b(x,\theta)$ and $c(x,\theta)$ cannot vanish simultaneously. Can you estimate them? Now establish (9.13).

7. It is clear that $\psi(x,\theta)$ contains a certain arbitrariness. Show that it is indeed possible to define it in such a way that it is continuously differentiable in both x and θ and so that $\psi(x,\theta)$ approaches zero as x approaches infinity.

8. Establish (9.18).

9. Using the transformation (9.26) verify (9.27) and (9.28).

10. Consider the problem (9.1a, b, c) with $d_1=0$, $d_2=1$, and with

$$f(t)=\begin{cases} 2, & 0\leqslant t\leqslant t_0,\\ 0, & t>t_0.\end{cases}$$

This problem has an analytical solution. Find it and hence determine the phase shift analytically. Now compute the phase shift numerically for various values of t_0 using both the very straightforward method of integrating "far enough out in t" (see Section 2) and by direct use of Theorem 9.2. Compare the computed results with each other and with the analytical answer. Try to obtain some idea of the relative efficacy of the two computational methods.

11. Continue the type of study initiated in Problem 10 by using functions f of your own choice. Resort to purely numerical answers when analytical solutions are not available. (They seldom are!)

12. Prove Lemma 9.4.

13. Carry out the derivation of Eq. (9.46).

14. Equation (9.57) is rejected in the text as a computational device on the grounds that the parameter λ enters into so many of the functions. Program Eq. (9.57) and use it to compute the results obtained in Examples 1 and 2. Now numerically solve these same two examples using the method of the text. Compare the speed, complexities, and efficiencies of the two programs.

15. Study the analyticity properties of the function $\tilde{b}$ and verify the statements made at the end of Section 11. Compare the difficulty of the proofs you have obtained with those commonly given (see [6]).

16. Compute the eigenvalues obtained in Examples 1 and 2 using the method of Chapter 8 by replacing the condition at infinity by the condition that the eigenfunctions actually vanish for some sufficiently large t values. Compare results, programs, efficiency, and so on.

17. Consider the equation $y''-(1+ze^{-at})y=0$, $y(0)=c_1$, $y'(0)=c_2$, $a>0$. Set $f(c_1,c_2,z)=\lim_{t\to\infty} y(t)e^{-t}$. Obtain a partial differential equation satisfied by f.

18. Obtain a power series expansion of the function f in the above problem in terms of z.

19.* The matrix analogue of the equation appearing in Problem 17 is just $x''-(A+ze^{-Bt})x=0$ where A and B are constant square matrices, x is a vector and z is a scalar parameter. Study this equation using the method of the preceding problems.

20.* Investigate the equations

$$u''+(1+ze^{-at})u=0,$$

$$w''\pm\left(1+\frac{z}{(t+a)^2}\right)w=0$$

by the methods of this chapter or by the device described in Problem 17. (See R. Bellman, "On Asymptotic Behavior of Solutions of Second-Order Differential Equations," *Quart. Appl. Math.* **20**, 1963, 385–387.)

REFERENCES

1. G. M. Wing, "Invariant Imbedding and the Asymptotic Behavior of Solutions to Initial Value Problems," *J. Math. Anal. Appl.* **9**, 1964, 85–98.
2. R. C. Allen, Jr. and G. M. Wing, "A Method for Computing Eigenvalues of Certain Schrödinger-like Equations," *J. Math. Anal. Appl.* **15**, 1966, 340–354.
3. F. A. Calogero, "A Novel Approach to Elementary Scattering Theory," *Nuovo Cimento*, **27**, 1963, 261–302.
4. W. A. Beyer, "Asymptotic Phase and Amplitude for a Modified Coulomb Potential in Scattering Theory: An Application of Invariant Imbedding," *J. Math. Anal. Appl.* **13**, 1966, 348–360.
5. F. G. Hagin, "Some Asymptotic Behavior Results for Initial Value Problems: An Application of Invariant Imbedding," *J. Math. Anal. Appl.* **20**, 1967, 540–564.
6. E. C. Titchmarsh, *Eigenfunction Expansions Associated with Second Order Differential Equations*, Oxford University Press (Clarendon), 1946.
7. F. G. Hagin, "Computation of Eigenvalues for Second Order Differential Equations Using Invariant Imbedding Techniques," *J. Comp. Phys.* **3**, 1968, 46–57.

10

APPLICATIONS TO EQUATIONS WITH PERIODIC COEFFICIENTS

1. INTRODUCTION

The work we shall describe in this chapter was originally undertaken to determine if the method of invariant imbedding had any especially useful properties that could be employed in the study of particle transport or wave propagation through periodic media, such as materials with crystal-like structure [1]. It was discovered that the technique is particularly powerful in such considerations. By combining properly the differential equation approach with some of the ideas on difference equations mentioned in Chapter 5, we effectively treated the physical questions at hand. In addition, there emerged a numerical method that seems extremely powerful when applied to any linear two-point boundary value problem with periodic coefficients.

Original investigation was confined to the one-state case, corresponding to scalar u and v equations of the kind encountered early in the first chapter of this book. The condition that the coefficients of these equations be periodic functions, all with the same period, was, of course, added to mimic the physical situation. Results—both analytic and numerical—were so interesting that the entire investigation was extended to the n state situation, also with success. In this chapter we shall concentrate on the scalar case, since it amply illustrates the ideas involved, merely mentioning very briefly the matrix analog to indicate the kind of difficulties that arise.

Finally, it was noted that some of the identities generated in the analysis were rather direct generalizations of classical trigonometric identities. In fact, by appropriate choice of our differential equations we were able to obtain the well-known identities for the trigonometric functions and for the hyperbolic functions. Later, careful investigation of these formulas revealed that it was only by chance that they were the outcome of investigations involving periodic media—they actually have nothing to do with the periodicity of the coefficients in the differential equations under study and are direct products of the functional relationships discussed in Chapter 3. Some of these more general relationships are found in Section 4 of that chapter. Indeed, most of the identities obtained in the present chapter can be derived directly from the formulas of Chapter 3. We choose to give basically independent derivations, however. These matters are further illuminated in the problem section.

2. STATEMENT OF THE PROBLEM

We consider the usual system of ordinary differential equations

$$\frac{du}{dz} = a(z)u(z) + b(z)v(z), \tag{10.1a}$$

$$-\frac{dv}{dz} = c(z)u(z) + d(z)v(z), \qquad x \leqslant z \leqslant y, \tag{10.1b}$$

$$u(x) = u_x, \qquad v(y) = v_y, \tag{10.1c}$$

where the functions $a(z)$, $b(z)$, $c(z)$, and $d(z)$ are periodic in z with period p. There will be no loss of generality if we take $p=1$. For convenience, we shall also suppose these functions continuous, although that condition may be somewhat relaxed.

Without further comment we shall always assume that we are working on a basic interval $\bar{x} \leqslant z \leqslant \bar{y}$ such that for any x and y, $\bar{x} \leqslant x < y \leqslant \bar{y}$, and any u_x and v_y the two-point boundary value problem (10.1) has a unique solution. Finally, we shall require $y - x > 1$. If this condition does not hold then the periodicity of the coefficients plays no role. Under these hypotheses all of the subsequent manipulations, while they may appear to be formal at times, are legitimate. We shall often leave to the problem section the proofs that our results can indeed be rigorously established.

We now seek the value of $u(z)$ and $v(z)$ at any point in the interval $x \leqslant z \leqslant y$, with the additional stipulation that as much use as possible be made of the assumed periodicity of $a(z), \ldots, d(z)$. We shall, of course, eventually employ Eqs. (3.4).

3. THE DIFFERENTIAL EQUATIONS OF INVARIANT IMBEDDING OVER ONE PERIOD

So long as we are working over an interval of one period or less we must simply resort to the kinds of differential equations that have arisen elsewhere in this book. We refer especially to Chapter 3 and to problems of that chapter. For convenience we list below differential equations that are pertinent at this point:

$$\frac{d}{dz}r_r(x,z)=b(z)+[a(z)+d(z)]r_r(x,z)+c(z)r_r^2(x,z), \quad (10.2a)$$

$$r_r(x,x)=0; \quad (10.2b)$$

$$\frac{d}{dz}t_r(x,z)=[d(z)+c(z)r_r(x,z)]t_r(x,z), \quad (10.3a)$$

$$t_r(x,x)=1; \quad (10.3b)$$

$$\frac{d}{dz}t_l(x,z)=[a(z)+c(z)r_r(x,z)]t_l(x,z), \quad (10.4a)$$

$$t_l(x,x)=1; \quad (10.4b)$$

$$\frac{d}{dz}r_l(x,z)=c(z)t_l(x,z)t_r(x,z), \quad (10.5a)$$

$$r_l(x,x)=0. \quad (10.5b)$$

It should be observed that this system is complete and that the functions r and t may be computed numerically, or, in rare cases, analytically. Furthermore, there is no prohibition against using the above equations over an interval of length greater than unity (one period). We simply choose not to do so.

4. DIFFERENCE EQUATIONS OVER AN INTEGRAL NUMBER OF PERIODS

We first observe that because of the assumed periodicity we have

$$r_r(z,z+1)=r_r(z+n,\, z+n+1) \quad (10.6)$$

for any integer n. A similar relationship holds for r_l and for the t functions.

Henceforth we write

$$r_r(z,z+1)=\rho_r(z), \qquad t_r(z,z+1)=\tau_r(z),$$

$$r_l(z,z+1)=\rho_l(z), \qquad t_l(z,z+1)=\tau_l(z). \tag{10.7}$$

Now let $y-x=n$, n a positive integer. Suppose that through the integration of the equations of Section 3 the r and t functions are already known over one period. We propose to derive equations for $r(x,x+k)$ and $t(x,x+k)$ for $k=2, 3,\dots,n$ through the use of difference equations. The reader will note that these difference equations have points of contact with those found in Chapter 5. However, our approach is rather different, as mentioned in that chapter.

We turn to the basic equations (3.3) and carefully select the independent variables:

$$u(x+k)=u(x+k-1)t_l(x+k-1,\, x+k)+v(x+k)r_r(x+k-1,\, x+k), \tag{10.8a}$$

$$v(x+k)=u(x+k)r_l(x+k,\, x+k+1)+v(x+k+1)t_r(x+k,x+k+1). \tag{10.8b}$$

As a matter of notational convenience we set

$$u(x+k)=\hat{u}(k), \qquad v(x+k)=\hat{v}(k). \tag{10.9}$$

Using (10.7) we then rewrite (10.8) as

$$\hat{u}(k)=\hat{u}(k-1)\tau_l(x)+\hat{v}(k)\rho_r(x), \tag{10.10a}$$

$$\hat{v}(k)=\hat{u}(k)\rho_l(x)+\hat{v}(k+1)\tau_r(x). \tag{10.10b}$$

After some manipulation these may be put in the form

$$\hat{u}(k)=\hat{a}(x)\hat{u}(k-1)+\hat{b}(x)\hat{v}(k-1), \tag{10.11a}$$

$$-\hat{v}(k)=\hat{c}(x)\hat{u}(k-1)+\hat{d}(x)\hat{v}(k-1), \tag{10.11b}$$

where

$$\hat{a}(x) = \frac{[\tau_l(x)\tau_r(x) - \rho_l(x)\rho_r(x)]}{\tau_r(x)}, \tag{10.12a}$$

$$\hat{b}(x) = \frac{\rho_r(x)}{\tau_r(x)}, \tag{10.12b}$$

$$\hat{c}(x) = \frac{\rho_l(x)}{\tau_r(x)}, \tag{10.12c}$$

$$\hat{d}(x) = \frac{-1}{\tau_r(x)}. \tag{10.12d}$$

We leave the fact that the denominator in these expressions cannot vanish as a problem for the reader (see Problem 4).

As yet we have not mentioned boundary conditions for (10.11). Our choice will, of course, be made in such a way as to make our current analysis applicable to the basic problem of this chapter. We require

$$\hat{u}(0) = u(x) = 0, \tag{10.13a}$$

$$\hat{v}(n) = v(y) = 1. \tag{10.13b}$$

By the definition of the r and t functions

$$\hat{u}(n) = u(y) = r_r(x, x+n), \tag{10.14a}$$

$$\hat{v}(0) = v(x) = t_r(x, x+n). \tag{10.14b}$$

The work of Chapter 3 suggests strongly that we define a new quantity $\hat{r}_r(k)$ by

$$\hat{u}(k) = \hat{r}_r(k)\hat{v}(k). \tag{10.15}$$

Equations (10.11) and a bit of manipulation then yield

$$\hat{r}_r(k) = -\frac{\hat{a}(x)\hat{r}_r(k-1) + \hat{b}(x)}{\hat{c}(x)\hat{r}_r(k-1) + \hat{d}(x)}, \tag{10.16a}$$

and, of course, we also have

$$\hat{r}_r(0) = 0. \tag{10.16b}$$

An analytic solution for the system (10.16) may be found (see [1] and Problem 5), and the question of the possible vanishing of the denominator in (10.16a) for some k, $k=2, 3,\ldots,n$ may also be dispensed with (see Problem 6). In practice it is usually much easier to compute directly from Eq. (10.16) (which is, of course, really a recursion formula for $\hat{r}_r$) than to make use of the analytic results.

From (10.11b) we discover

$$\frac{\hat{v}(k)}{\hat{v}(k-1)} = -\left[\hat{c}(x)\hat{r}_r(k-1)+\hat{d}(x)\right], \tag{10.17}$$

and this yields

$$\hat{v}(0)=\frac{(-1)^n\hat{v}(n)}{\prod_{k=0}^{n-1}\left[\hat{c}(x)\hat{r}_r(k)+\hat{d}(x)\right]}. \tag{10.18}$$

It is now appropriate to put our results in terms of the original r and t functions. Tracing through our definitions we find

$$r_r(x,x+n)=\hat{r}_r(n) \tag{10.19}$$

and so

$$r_r(x,x+j)=-\frac{\hat{a}(x)r_r(x,x+j-1)+\hat{b}(x)}{\hat{c}(x)r_r(x,x+j-1)+\hat{d}(x)}, \qquad j=1,2,\ldots,n, \tag{10.20a}$$

with

$$r_r(x,x)=0. \tag{10.20b}$$

Similarly

$$t_r(x,x+n)=\frac{(-1)^n}{\prod_{k=0}^{n-1}\left[\hat{c}(x)r_r(x,x+k)+\hat{d}(x)\right]}, \tag{10.21a}$$

$$t_r(x,x)=1. \tag{10.21b}$$

We have thus accomplished our goal of finding two of the r and t functions at arbitrary "period points" by use merely of difference equations.

It is next necessary to obtain similar formulas for r_l and t_l. Basically, the device employed here is to use y as the reference point instead of x. We write the equations analogous to (10.8) in the form

$$u(y-j)=u(y-j-1)t_l(y-j-1,y-j)+v(y-j)r_r(y-j-1,y-j), \tag{10.22a}$$

$$v(y-j)=u(y-j)r_l(y-j,y-j+1)+v(y-j+1)t_r(y-j,y-j+1). \tag{10.22b}$$

Now a set of manipulations that parallels that used for the derivation of (10.20) and (10.21) eventually yields (see [1]; also Problem 7)

$$r_l(x,x+j)=-\frac{\tilde{c}(x)+\tilde{d}(x)r_l(x,x+j-1)}{\tilde{a}(x)+\tilde{b}(x)r_l(x,x+j-1)}, \qquad j=1,2,\ldots,n, \tag{10.23a}$$

$$r_l(x,x)=0, \tag{10.23b}$$

$$t_l(x,x+n)=\frac{1}{\prod_{k=0}^{n-1}\left[\tilde{a}(x)+\tilde{b}(x)r_l(x,x+k)\right]}, \tag{10.24a}$$

$$t_l(x,x)=1, \tag{10.24b}$$

where

$$\tilde{a}(x)=\frac{1}{\tau_l(x)}, \tag{10.25a}$$

$$\tilde{b}(x)=\frac{-\rho_r(x)}{\tau_l(x)}, \tag{10.25b}$$

$$\tilde{c}(x)=\frac{-\rho_l(x)}{\tau_l(x)}, \tag{10.25c}$$

$$\tilde{d}(x)=\frac{[\rho_r(x)\rho_l(x)-\tau_r(x)\tau_l(x)]}{\tau_l(x)}. \tag{10.25d}$$

5. DIFFERENCE EQUATIONS OVER A NONINTEGRAL NUMBER OF PERIODS

We now investigate the possibility of deriving difference equations for the r and t functions when $y-x=n+w$, where n is a positive integer and $0<w<1$. Once again the fundamental system (3.3) plays the key role. In that set we choose $z_1=x$, $z_2=x+n+w$, $u(z_1)=0$, and $v(z_2)=1$ to obtain

$$u(z)=v(z)r_r(x,z), \tag{10.26a}$$

$$v(z)=u(z)r_l(z,x+n+w)+t_r(z,x+n+w), \tag{10.26b}$$

or

$$u(z)=\frac{r_r(x,z)t_r(z,x+n+w)}{1-r_r(x,z)r_l(z,x+n+w)}. \tag{10.27}$$

Now choose $z=z_1$ in (10.27) and $z=x+n+w$ in (3.3a):

$$u(x+n+w)=\frac{r_r(x,z_1)t_r(z_1,x+n+w)t_l(z_1,x+n+w)}{1-r_r(x,z_1)r_l(z_1,x+n+w)} + r_r(z,x+n+w)v(x+n+w). \tag{10.28}$$

But our choices of arguments have yielded

$$v(x+n+w)=1, \qquad u(x+n+w)=r_r(x,x+n+w),$$

so that (10.28) may be rewritten

$$r_r(x,x+n+w)=\frac{r_r(x,x+n)t_r(x+n,x+n+w)t_l(x+n,x+n+w)}{1-r_r(x,x+n)r_l(x+n,x+n+w)} + r_r(x+n,x+n+w). \tag{10.29}$$

Using the observation (10.6) we may simplify this to

$$r_r(x,x+n+w)=\left[r_r(x,x+n)\{t_r(x,x+w)t_l(x,x+w) - r_r(x,x+w)r_l(x,x+w)\}+r_r(x,x+w)\right] \times\left[1-r_r(x,x+n)r_l(x,x+w)\right]^{-1}. \tag{10.30}$$

Note that all quantities on the right side of the above equation are calculable either by use of the differential equations of Section 3 or from the difference equations of Section 4. Equation (10.30) is therefore the desired result for r_r.

Analogous reasoning (see Problem 8) produces

$$r_l(x,x+n+w)=\big[r_l(x,x+w)\{t_r(x,x+n)t_l(x,x+n)-r_r(x,x+n)r_l(x,x+n)\}+r_l(x,x+n)\big]\times[1-r_r(x,x+n)r_l(x,x+w)]^{-1}, \tag{10.31}$$

$$t_r(x,x+n+w)=\frac{t_r(x,x+n)t_r(x,x+w)}{1-r_r(x,x+n)r_l(x,x+w)}, \tag{10.32}$$

$$t_l(x,x+n+w)=\frac{t_l(x,x+n)t_l(x,x+w)}{1-r_r(x,x+n)r_l(x,x+w)}. \tag{10.33}$$

We leave the appropriate initial conditions to the reader.

6. THE "BACKWARDS" EQUATIONS

In one sense our work is done since enough information has been derived to make use of Eqs. (3.4) or (3.5) to find the values of u and v at any internal point, z. However, those equations will involve such expressions as $t_r(z,y)$, $r_l(z,y)$, etc., and we have always used the first argument in the r and t functions as the left end point, x. It may be argued that the equations we have already derived will suffice, since one may always consider z to be fixed and y to be variable. If one is interested in only one or two values of $u(z)$ and $v(z)$ this is a reasonable approach. However, a little thought reveals that if many z arguments are to be examined, then the method we have suggested is very wasteful of computer time. It is therefore desirable to find a set of equations analogous to those of the last three sections which produce the functions $r(z,y)$ and $t(z,y)$ directly. This is easy to do, but quite tedious. The equations resemble in form the ones we have already derived, and it does not seem at all valuable to write them down here. The reader who may be interested in making actual numerical calculations will find a complete set given in [1] (see also Problems 9 and 10). We refer to these equations as the "backwards" equations because one is working from the right-hand end point of the interval of interest to the left.

7. SOME NUMERICAL RESULTS

We shall present just two sets of numerical results obtained with the theory we have developed. For more extensive tables and further examples, the interested reader is referred to [1] (see also [2]). The examples were computed on an IBM 360/40 machine. A fourth order Runge–Kutta scheme was employed for the integration of the necessary differential equations, and double precision arithmetic was used. Elsewhere in the calculations single precision sufficed.

Example 1

We begin with the problem

$$\psi''(z)-\psi(z)=0, \qquad \psi(0)=1, \qquad \psi'(y)=e^{-y}, \qquad 0\leqslant z\leqslant y, \quad (10.34)$$

equivalent to the system

$$\frac{du}{dz}=v, \quad (10.35a)$$

$$-\frac{dv}{dz}=-u, \quad (10.35b)$$

$$u(0)=1, \qquad v(y)=-e^{-y}. \quad (10.35c)$$

This is a particularly interesting, although trivial, system since one of the fundamental solutions of the differential equation is $\psi(z)=e^{z}$. For large y this solution can be very troublesome numerically, even though the boundary conditions are chosen to suppress it completely. Indeed the solution to (10.35) is simply

$$u(z)=e^{-z}, \qquad v(z)=-e^{-z}. \quad (10.36)$$

Since the coefficients in (10.35a, b) are constant we are free to pick the period as we choose. [Recall that all formulas of this chapter were derived under the assumption that p, the period, was equal to unity. This was simply for convenience, and the necessary modifications for an arbitrary period are easily made. They are left to the reader (see Problem 11).] We rather arbitrarily chose $p=0.125$ in our calculations. Table 1 reveals the exceptional accuracy obtained. Note that at $y=100.0$ the difference equations have been employed almost 800 times, and that "lurking in the background" is a term of magnitude $e^{100} \approx 10^{43}$ which the method has virtually completely suppressed.

TABLE 1

y	Computed Value, $u(y/2)$	Correct Value, $e^{-\frac{1}{2}y}$
1.0	0.60653	0.60653
5.0	0.82086×10^{-1}	0.82085×10^{-1}
10.0	0.67382×10^{-2}	0.67379×10^{-2}
30.0	0.30594×10^{-6}	0.30590×10^{-6}
50.0	0.13891×10^{-10}	0.13888×10^{-10}
100.0	0.19295×10^{-21}	0.19287×10^{-21}

Example 2

Here we consider the system

$$u' = v, \tag{10.37a}$$

$$-v' = \frac{k\alpha\beta(\beta \sin \beta z + k \cos \beta z)u}{k(1+\alpha \sin \beta z) - \alpha\beta \cos \beta z} - \frac{[\alpha(\beta^2 + k^2) \sin \beta z + k^2]v}{k(1+\alpha \sin \beta z) - \alpha\beta \cos \beta z},$$

$$0 \leqslant z \leqslant y, \tag{10.37b}$$

$$u(0) = 1, \qquad v(y) = \alpha\beta \cos \beta y. \tag{10.37c}$$

Again the system (10.37a, b) has a fundamental solution $u(z) = e^{kz}$, but the boundary conditions have been so contrived that the solution to the full problem is actually just

$$u(z) = 1 + \alpha \sin \beta z, \qquad v(z) = \alpha\beta \cos \beta z. \tag{10.38}$$

For this problem the values $k = 1$, $\alpha = (\frac{1}{2}\pi)^2$, and $\beta = 2\pi$ were selected. Obviously, this choice yields $p = 1$. Table 2 gives a brief resume of some of the results.

It is always unwise to declare the superiority of a numerical method on the basis of its success in treating a few special cases, regardless of how ill-behaved those cases may be. However, we have yet to encounter a system of the form (10.1) in which the imbedding method has not given reasonable results. A careful analysis of the technique with a study of round-off errors, truncation errors, and so on would certainly be most valuable. This has not been undertaken.

TABLE 2

	Computed Value		Correct Value	
y	$u(y/2)$	$v(y/2)$	$1+\alpha\beta\sin(\beta y/2)$	$\alpha\beta\cos(\beta y/2)$
1.0	1.00000	−0.15915	1.00000	−0.15915
10.0	1.00000	+0.15916	1.00000	+0.15915
100.0	1.00003	−0.15916	1.00000	−0.15915
300.0	1.00009	+0.15917	1.00000	+0.15915
500.0	1.00014	+0.15918	1.00000	+0.15915

8. THE METHOD OF DOUBLING

Let us return to Section 5. There the requirement was made that w lie strictly between zero and unity. This was solely because our major interest was in deriving r and t equations for a nonintegral number of periods. A rereading of the material of that section will reveal that the restriction was by no means necessary, and so long as $x+n+w \leqslant \bar{y}$ the results are valid. Suppose in (10.30) we pick $n=1$ and $w=1$. Then

$$\begin{aligned} r_r(x,x+2)=\big[r_r(x,x+1)\{&t_r(x,x+1)t_l(x,x+1) \\ &-r_r(x,x+1)r_l(x,x+1)\}+r_r(x,x+1)\big] \\ &\times\big[1-r_r(x,x+1)r_l(x,x+1)\big]^{-1}. \end{aligned} \tag{10.39}$$

Similarly, from (10.32) we obtain

$$t_r(x,x+2)=\frac{t_r^2(x,x+1)}{1-r_r(x,x+1)r_l(x,x+1)}. \tag{10.40}$$

Clearly, values for $r_l(x,x+2)$ and $t_l(x,x+2)$ may be obtained in the same way. Repeating the procedure with $n=w=2$ we can obtain values for the r and t functions at $x+4$, and, by iteration at $x+2^k$, $k=3, 4,\ldots,$the only restriction being that $x+2^k$ not exceed $\bar{y}$.

The process we have described is a generalization of the "method of doubling" first used in transport problems with constant parameters by van de Hulst [3]. (See Chapter 3, Section 4.) It clearly possesses advantages in time and effort saved when one is examining a system that is many periods "thick." There is also very likely a considerable improvement in accuracy obtained. For example, a system in which $y-x=1024$ requires only ten uses of the recursion formulas, allowing much less opportunity for build up of round off error than is the case when Section 4 is used directly.

A similar device can be used with respect to the "backwards equations" which we have not written down. Obviously judicious use of the method can often be used to compute systems of any size more efficiently than the straightforward approach we have been using. For example, the value $y=500$ in Table 2 could have been reached by "doubling ahead" to $512=2^9$, then "doubling back" to 504 in three steps ($2^3=8$), and finally "doubling back" to 500. This capacity has not yet been built into our experimental programs, however, and all examples of the preceding section were computed without the "doubling" device.

9. TRIGONOMETRY REVISITED

The reader may very well have noticed that many of the recursion formulas we have derived are somewhat reminiscent of trigonometric identities. Let us consider the special problem

$$\frac{du}{dz}=v, \qquad (10.41a)$$

$$0 \leqslant z < \frac{\pi}{2}.$$

$$-\frac{dv}{dz}=u. \qquad (10.41b)$$

The r and t functions for (10.41) are trivially computed:

$$r_r(0,z)=r_l(0,z)=\tan z$$

$$t_r(0,z)=t_l(0,z)=\sec z. \qquad (10.42)$$

We also note that the period p in the system (10.41) is quite arbitrary, since the coefficients are constant. If we use (10.30) and (10.32) rewritten for the case of arbitrary p, substitute (10.42) and do a little algebra (see Problem 13) we obtain

$$\tan(np+w)=\frac{\tan(np)+\tan w}{1-\tan(np)\tan w}, \qquad (10.43a)$$

$$\frac{1}{\cos(np+w)}=\frac{1}{\cos(np)\cos w-\sin(np)\sin w}, \qquad (10.43b)$$

provided $0 \leqslant np+w<\pi/2$. We have thus obtained two of the standard trigonometric identities, although subject to conditions on their arguments far more severe than are imposed in standard theory. Our derivation is thus only partially satisfactory, and leads to some interesting questions which will be mentioned in the final section.

By examining the set

$$\frac{du}{dz} = v, \qquad (10.44a)$$

$$0 \leqslant z < \infty,$$

$$\frac{dv}{dz} = u, \qquad (10.44b)$$

we can similarly obtain identities for the hyperbolic functions, this time subject only to the requirement that all arguments merely be positive.

As mentioned in the introduction to this chapter, the results we have just derived using our theory can actually be obtained without the assumption of periodicity, and the unusually severe restrictions on the arguments in the examples discussed may be dropped completely. Indeed, this program was initiated in Chapter 3. As noted earlier, many of the results we have presented could have been obtained by pursuing the methods of that chapter (see the problem section). We have chosen to use a more direct approach making full use of the periodicity.

10. SUMMARY

In this chapter we have shown how a judicious combination of the ideas of invariant imbedding as applied to differential equations and to difference equations can lead to a very efficient algorithm for the solution of two-point boundary value problems in which the coefficients are periodic functions of the independent variable. Attention has been confined to the scalar case for simplicity. Basically the matrix case follows the same pattern, but care must be exercised in writing the various final formulas so as not to violate noncommutativity principles. In addition, there arise some rather interesting questions concerning the existence of certain inverse matrices and the like. The reader is referred to [2] where the matter is covered in detail.

There is, of course, a classical theory due to Floquet for the study of periodic problems. We have chosen to make no use of this theory. It seems likely that the Floquet theory can be obtained from our results, but no attempt has been made to do this. Moreover, this classical theory concerns itself mainly with behavior of fundamental solutions to periodic problems, rather than to two-point boundary questions. As we have seen in some of the numerical examples, it is the very ability of the foregoing method to suppress certain of these fundamental solutions that makes this algorithm of particular interest and value.

PROBLEMS

1. Show that the restriction that the coefficients in (10.1a, b) be continuous can easily be relaxed to piecewise continuity. Investigate the possibility of allowing somewhat more serious ill-behavior in the coefficients.

2. Verify that the seemingly formal manipulations of the entire chapter are indeed valid under the restrictions imposed in the second paragraph of Section 2. Discuss these restrictions and their relation to eigenlengths and eigenvalues.

3. Derive Eqs. (10.2)–(10.5). (See Problem 10 of Chapter 3.)

4. Verify the formulas for $\hat{a}(x), \hat{b}(x)$, and so on, and show that the denominator cannot be zero.

5. Solve the system of difference equations (10.16) analytically, taking note of any special cases. Compare the possible usefulness of this analytic solution with the numerical solution obtained in an obvious way from (10.16). Program the problem and run some examples.

6. Prove that the denominator of (10.16a) cannot vanish.

7. Obtain Eqs. (10.23)–(10.25).

8. Derive Eqs. (10.31)–(10.33) using the methods of this chapter.

9. Write down in detail all the "backwards" equations referred to in Section 6.

10. Use the results and ideas of Chapter 3 [for example, Eqs. (3.8) and (3.11)] to obtain all of the difference equations used in this chapter for the various r and t functions. Generalize all of these to the matrix case. Now generalize all results of this chapter to the matrix case.

11. Revise the various formulas of this chapter to allow for a general period p instead of the unit period assumed.

12.* Write a computer program that makes full use of "doubling forward" and "doubling backward." (See the last paragraph of Section 8.) Observe that an interesting number–theoretic problem is involved: namely, what is the most efficient way of reaching the integer m if one can "double ahead" from unity and can also "double back" from values greater than m?

13. Discuss the derivation of the trigonometric identities from the viewpoint of Chapter 3. By starting with equations with constant coefficients different from (10.41) obtain other identities involving trigonometric and hyperbolic functions. Note that the same ideas can be used on any set of equations with periodic coefficients. Try to obtain some identities involving the Mathieu functions, for example.

14.* The restriction imposed in the second paragraph of Section 2 has been vital to the analysis. Consider the possibility of removing this restriction in the following way. Suppose there is a point z_1, $\bar{x} < z_1 < \bar{y}$, such that if $x = z_1$ or $y = z_1$, the corresponding problem cannot be solved for all possible u_x and v_y. For convenience, let z_1 be the only such point. Show that it is possible, through the use of the ideas of Chapter 3, to write valid identities for r and t functions which, in effect, "pull" these functions across the point z_1. (Note that the usual trigonometric identities do just this. The tangent function may be thought of as the solution $u(z)$ to (10.41). This provides a definition in case $0 \leqslant z < \pi/2$. The identity (10.43a) "pulls" the tangent function across the point $z = \pi/2$.)

15.* The ideas suggested in Problem 14 really have relatively little to do with the periodicity features of the current study. Obtain a general theory for handling problems that are ill-behaved in the way described. Observe that the assumption that there is only one "bad" point z_1 may be dropped quite easily. Connect your results with the problem of eigenvalues and eigenlengths. (See R. C. Allen, Jr. and G. M. Wing. "An Invariant Imbedding Algorithm for the Solution of Inhomogeneous Linear Two-Point Boundary Value Problems," *J. Comput. Phys.* **14**, 1974, pp. 40–58.)

16. Show that the results of this chapter may be extended to systems of m equations, m even or odd, with q values assigned at the left end and s values at the right, $q+s=m$, but q not necessarily equal to s. (See Problem 12, Chapter 3, and Problem 8, Chapter 2.)

17.* Establish the fact that a nonsingular matrix can always be represented as an exponential. That is, if $\det A \neq 0$, then there is a matrix B such that $A = e^B$. Proceed in stages as follows:

a. The result is true if A is diagonal.

b. The result is true if A can be diagonalized; that is, if there is a T such that TAT^{-1} is diagonal.

c. The result is true if A is upper triangular. (*Hint.* Do the 2×2 case first.)

d. Every nonsingular matrix may be converted into an upper triangular matrix by a similarity transformation.

(See R. Bellman, *Matrix Analysis*, 2nd ed., McGraw-Hill, New York, 1972.)

18.* With the use of the result obtained in Problem 17, establish the theorem of Floquet: Any solution of the matrix equation $X' = P(t)X$ where $P(t)$ is periodic with period unity can be written in the form $X(t) = Q(t)e^{Bt}$ where $Q(t)$ is periodic with unit period and B is constant. Do this by noting that if $X(t)$ satisfies the equation then so does $X(t+1)$.

19.* Can the methods of this chapter be used to establish the Floquet theorem, at least in the second order case?

20. Consider the matrix equation $X' = (P_1(t) + P_2(t+\theta))X$, $X(0) = I$, where $P_1(t) = P_1(t+1)$, $P_2(t) = P_2(t+\lambda)$, and λ is irrational. Denote the solution by $X(t,\theta)$. Prove that $X(t+1,\theta) = X(t,\theta+1)X(1,\theta)$. Thus show that $X(t+n,\theta) = X(t,\theta+n)$ $X(1,\theta+n-1)X(1,\theta+n-2)\cdots X(1,\theta)$.

21.* Combine the results of Problems 17 and 20 to show that

$$\prod_{k=0}^{n} X(1,\theta+n-k) = \prod_{k=0}^{n} [\exp B(\theta+n-k)],$$

where $X(1,\theta) = e^{B(\theta)}$. Does an expression of this form have any simple asymptotic behavior as $n \to \infty$?

22.* Is the invariant imbedding method of any value in studying equations of the type occurring in Problem 20?

23. The equation of Mathieu, $u'' + (a + b \sin t)u = 0$ is covered by the treatment of this chapter. So is the equation $y'' + (a + b \operatorname{sn}(t))y = 0$ where $\operatorname{sn}(t)$ is the (elliptic) sine-amplitude function. A classical way of treating the Mathieu equation is to note that as the parameter k, which occurs in the sn function, approaches zero the sine-amplitude function approaches the ordinary sine function. Thus an explicit solution of the Mathieu equation can be obtained in terms of the limit of appropriate solutions y, which are actually doubly periodic functions. This yields both an analytical and computational device for the study of the Mathieu equation.

Make some numerical investigations of these equations using the methods of this chapter.

REFERENCES

1. R. C. Allen, Jr. and G. M. Wing, "A Numerical Algorithm Suggested by Problems of Transport in Periodic Media," *J. Math. Anal. Appl.* **29**, 1970, 141–157.
2. R. C. Allen, Jr., J. W. Burgmeier, P. Mundorff, and G. M. Wing, "A Numerical Algorithm Suggested by Problems of Transport in Periodic Media: The Matrix Case," *J. Math. Anal. Appl.* **37**, 1972, 725–740.
3. H. C. Van de Hulst, *A New Look at Multiple Scattering*, NASA Institute for Space Studies, Goddard Space Flight Center, January 1963.

11

TRANSPORT THEORY AND RADIATIVE TRANSFER

1. INTRODUCTION

As we indicated earlier in this book, the great stimulus to the study of invariant imbedding came through the work of Ambarzumian and Chandrasekhar, whose interests were largely in astrophysical applications. Specifically, they used the basic concepts in the investigation of problems of radiative transfer. Subsequently applications to transport theory were made, leading, for example, to reactor shielding studies. We have made reference to very simple transport models in our discussions in the preceding chapters. Except for technical details of interest primarily to the physicist, engineer, astronomer, and so on, there is no essential difference between transport theory and radiative transfer, with one basic exception which we shall discuss in the next paragraph. Radiative transfer has classically been the term used for the study of the "flow" of light particles —that is, photons—in a medium, although the same expression has been employed in the study of X-ray phenomena. Here the distinction is primarily one of the frequency of the radiation under study. The term "transport theory" is generally used by the neutron physicist, but it may also be used in describing the movement of beta rays, gamma rays, and the like, in a medium. It is at once clear that there are regions of physical overlap within which either terminology might be applied. Often in radiative transfer problems one tends to think in terms of radiation and frequencies, while in transport problems one conceives of particles and energies. Because of the duality that exists between particle phenomena and wave phenomena at the microscopic level the difference is often just in the mode of thought and little else.

It was mentioned above that there is one major distinction between the two studies. The study of transport theory was greatly accelerated when the phenomenon of neutron induced fission was discovered. In this process —the essential ingredient of the atomic energy program that has had such a tremendous effect on so many aspects of life in the last several decades— a collision event may increase the total number of neutrons present in the system. (We have, of course, considered instances of this fission process in some of the idealized models of the earlier chapters.) In classical radiative transfer studies fissionlike processes did not occur. No more particles (photons) emerged from a collision than entered into the interaction, and frequently particles were lost through absorption processes. Hence, in the standard terminology, radiative transfer phenomena were at best conservative, and no opportunity for criticality existed. (We cautiously use the past tense in this discussion because evidences of photon induced fission have now been discovered; however, the process does not play an important role in radiative transfer studies.)

It is unfortunate that there are these two ways of referring to physical processes that are so similar. There has naturally resulted a special vocabulary used by the practitioners in each field, and this has led to further difficulties in communication. A few books (see, e.g., [1]) on one subject or the other provide a glossary of equivalent terms in the two areas. In keeping with our terminology of the earlier chapters (and the principal backgrounds of the authors) we shall use mainly the vocabulary of transport theory and refer the reader whose primary interest is in radiative transfer to one of the books noted above for suitable translation of terms.

2. THE LINEARIZED BOLTZMANN EQUATION

We consider particles in three-dimensional space. They are all of the same kind, but do not necessarily have the same velocities. Moreover, they are moving within a medium whose boundary and physical properties are completely specified. This region will be denoted by X. As in our simple models of earlier chapters, the particles do not interact with each other, but they can interact with the medium. We may consider such an interaction as involving a collision of the particle with a nucleus of the material of which the medium is constituted. The probability of such a collision is known, and may depend on particle location, velocity, time, and so on. The expected outcome of such a collision interaction is also known. The colliding particle always disappears, and new particles may emerge at the collision site (but only there) and they emerge instantaneously after the interaction. Their velocity distribution is given, and may depend on location, the velocity of the original particle, the time, and, of course, the type of material of which the medium is formed. We further suppose that a collision interaction in no way affects the medium itself. Finally, there may exist within the medium and on its surface spontaneous sources of particles. These are completely prescribed functions of position and velocity and of time. They in no way depend on the existing particle distribution in the medium.

The principle problem of transport theory as we shall view the subject is the following: Given the situation described in the previous paragraph, and information about what becomes of a particle when it emerges from the region X, describe the expected number of particles in the medium at any time as a function of position and velocity, assuming that such information is known at time zero.

Since the major topic of this book is the method of invariant imbedding and its various applications, it is not appropriate to get too deeply involved with the study of transport theory per se. We shall give one derivation of the basic governing equation, which is a linearized form of the famous equation of Boltzmann, provide a brief discussion of some of the boundary and initial conditions that are most often encountered, and then specialize the systems under study in such a manner that progress with the imbedding method can be made effectively. The reader who does not already have some knowledge of transport theory is referred to one of the many treatises on that subject [2–4]. For those readers whose previous contact has been with the field of radiative transfer we recommend such a glossary of equivalent terms as was discussed in the previous section.

We begin by defining the function $N(x,y,z,v_x,v_y,v_z,t)$:

$$N(x,y,z,v_x,v_y,v_z,t) = \text{expected density of particles at time } t \text{ at position } (x,y,z) \text{ in } X \text{ and having velocity components } v_x, v_y, v_z, \text{ elements of a set } V. \tag{11.1}$$

As a matter of notational convenience one frequently writes $N = N(\mathbf{r}, \mathbf{v}, t)$ where $\mathbf{r}$ denotes the position vector and $\mathbf{v}$ the velocity vector. Sometimes it is desirable to write $v_x = \Omega_x v$, $v_y = \Omega_y v$, and $v_z = \Omega_z v$, where Ω_x, Ω_y, and Ω_z are the direction cosines of the velocity vector and v is the particle speed. In this case one often writes $N = N(x,y,z,\Omega_x,\Omega_y,\Omega_z,v,t)$. Other notational modifications are rather obvious. Although the use of the same letter N followed by a variety of different arguments can usually be considered as a mathematical abomination, we shall take care in our development to clarify what the argument list is in any particular investigation.

Let us return for a moment to the basic definition (11.1). By *expected density* we mean that given a set $\tilde{X}$ in position space and a set $\tilde{V}$ in velocity space then the expected number of particles found in $\tilde{X}\times\tilde{V}$ at time t is given by

$$\int_{\tilde{X}\times\tilde{V}} N(\mathbf{r},\mathbf{v},t)\,d\mathbf{r}\,d\mathbf{v}. \tag{11.2}$$

That the word *expected* is needed is obvious from the stochastic aspects of the model described. We shall not dwell on such matters as the generality of the sets $\tilde{X}$ and $\tilde{V}$ allowed in (11.2) nor on the type of integration to be used. Such matters are discussed in great detail in [5]. For our purposes the Riemann integral will ordinarily suffice, and all functions encountered will be assumed at least piecewise continuous, unless otherwise noted.

As in Chapter 7, we shall use a simple Lagrangian approach for the derivation of the Boltzmann equation. We observe the "beam" of particles at time t and at position (x,y,z) traveling in direction Ω_x, Ω_y, and Ω_z, at speed v. With Ω_x, Ω_y, Ω_z, and v *fixed*, how does N vary as t changes? Obviously, we are talking about the total time derivative DN/Dt when only x, y, z (and, of course, t) are allowed to vary. Any such change in N can occur in only three possible ways:

1. Particles in the beam interact with the medium and therefore disappear from the beam;
2. Particles are added to the beam because of the presence of spontaneous sources;
3. Particles in other beams (that is with other velocities $\mathbf{v}'$) have collisions, and the new particles thus created contribute to the beam under investigation. [It should be noted that when event (1) above occurs, particles may also be created and move in the direction and speed of the original beam. We consider this event a special case of (3).]

In keeping with earlier definitions we define

$$\sigma(\mathbf{r},\mathbf{v},t)\Delta+o(\Delta)=\text{probability of a particle at position } \mathbf{r} \text{ and velocity } \mathbf{v} \text{ at time } t \text{ having a collision interaction in moving a distance } \Delta. \tag{11.3}$$

In the event of a collision at $\mathbf{r}$ at time t we assume that the expected density of particles produced in velocity state $\mathbf{v}$ when the colliding particle is moving in velocity state $\mathbf{v}'$ is $K(\mathbf{r},\mathbf{v}',\mathbf{v},t)$. That is, if $\tilde{V}$ is an arbitrary set in velocity space then the expected number of particles emergent from such a collision having velocity in the set $\tilde{V}$ is

$$\int_{\tilde{V}} K(\mathbf{r},\mathbf{v}',\mathbf{v},t)\,d\mathbf{v}. \tag{11.4}$$

Finally, we note that $\Delta = v\Delta t$, in an obvious notation. A simple conservation argument based on the fundamental events (1), (2), and (3) then gives

$$\begin{aligned} N(x+v_x\Delta t, y+v_y\Delta t, z+v_z\Delta t, \mathbf{v}, t+\Delta t) - N(x,y,z,\mathbf{v},t) \\ = -\sigma(\mathbf{r},\mathbf{v},t)v\Delta t N(\mathbf{r},\mathbf{v},t) + \int_{V'} \sigma(\mathbf{r},\mathbf{v}',t)v'\Delta t K(\mathbf{r},\mathbf{v}',\mathbf{v},t)N(\mathbf{r},\mathbf{v}',t)\,d\mathbf{v}' \\ + S(\mathbf{r},\mathbf{v},t)v\Delta t + o(\Delta), \end{aligned} \tag{11.5}$$

where $S(\mathbf{r},\mathbf{v},t)v\Delta t$ is the expected number of particles produced at $\mathbf{r}$ with velocity $\mathbf{v}$ in the time interval $(t, t+\Delta t)$ by virtue of the spontaneous sources. The integration is taken over all velocity space, V'. The usual assumptions concerning continuity and differentiability produce the relation

$$\begin{aligned} v_x\frac{\partial N}{\partial x} + v_y\frac{\partial N}{\partial y} + v_z\frac{\partial N}{\partial z} + \frac{\partial N}{\partial t} = \mathbf{v}\cdot \operatorname{grad} N + \frac{\partial N}{\partial t} \\ = \frac{DN}{Dt} = -\sigma(\mathbf{r},\mathbf{v},t)vN(\mathbf{r},\mathbf{v},t) + \int_{V'} \sigma(\mathbf{r},\mathbf{v}',t)v'K(\mathbf{r},\mathbf{v}',\mathbf{v},t)N(\mathbf{r},\mathbf{v}',t)\,d\mathbf{v}' \\ + vS(\mathbf{r},\mathbf{v},t). \end{aligned} \tag{11.6}$$

This is the fundamental equation of transport theory.

3. SOME REMARKS ON SECTIONS 1 AND 2

Various questions concerning the validity, both mathematical and physical, of the derivation of the foregoing basic equation may legitimately be asked. For example, it is not always true that the particles produced in a collision interaction all appear "spontaneously." In fact, in problems involving nuclear reactors, the "delayed" neutrons—those which do not appear until several milliseconds or more after the collision takes place—can be quite important. Again, the assumption that the *medium* is unchanged by the transport process is not strictly valid. A fission, for example, results in the destruction of the nucleus involved. Over a considerable length of time, where the word "considerable" is relative and may mean a few microseconds in the case of an explosive reaction and a few months in the case of a reactor, the medium definitely does change character. Or there may be other processes taking place, apart from the transport phenomena, which alter the medium. This can be the case in stars, for example. To include all such physical events would so complicate equation (11.6) as to make it virtually impossible to do any analysis other than strictly numerical, if even that. It must again be stressed that we are really dealing with an idealized model.

From the mathematical viewpoint, some of our standard assumptions concerning continuity are rather restrictive. For instance, it would seem desirable to be able to derive the equations of Chapter 1 (see (1.8) and (1.9)) as special cases of (11.6). This can be done if we relax the continuity assumptions, and go so far as to allow the employment of the Dirac delta function. (Equivalently, we might employ Stieltjes integrals or distributions instead of the simpler integrals we have been using.)

Let us pursue this subject. Suppose the medium under consideration is simply an infinite slab extending from $z=0$ to $z=a$. (We presently avoid using the symbol x as the right coordinate so as not to encounter confusion with the notation of this chapter.) Assume, moreover, that particles are capable of moving in only *two* directions—to the left and to the right—and that they all have the same speed, which may be taken as unity. All parameters of the medium depend only on the coordinate z, and the entire process is time-independent. Obviously Ω_z can take on only two values,

$$\Omega_z = +1, \qquad \Omega_z = -1. \tag{11.7}$$

We choose to write

$$N(z,+1) = u(z), \qquad N(z,-1) = v(z), \tag{11.8}$$

and define

$$\begin{aligned} K(z,v',-1) &= \delta(v'+1)f(z) + \delta(v'-1)b(z), \\ K(z,v',+1) &= \delta(v'+1)b(z) + \delta(v'-1)f(z). \end{aligned} \tag{11.9}$$

If we now notice that V' consists of just the number pair $(-1,+1)$ then substitution in (11.6) formally yields the two equations

$$\begin{aligned} \frac{du}{dz} &= -\sigma(z)u(z) + \{f(z)u(z) + b(z)v(z)\}\sigma(z), \\ -\frac{dv}{dz} &= -\sigma(z)v(z) + \{b(z)u(z) + f(z)v(z)\}\sigma(z), \end{aligned} \tag{11.10}$$

provided no extraneous sources are present. This is in agreement with (1.8) and (1.9).

It is interesting to note that while (11.10) agrees with the earlier equations, a few matters must still be reconciled. In Chapter 1, $u(z)$ and $v(z)$ were defined as numbers of particles passing a point; here they are densities. Also, in the early treatment we dealt with a "rod" geometry; now we are speaking of a slab. We leave to the reader the question of clarifying these matters (Problem 3). It may also be noted that had the time-dependence not been removed Eq. (11.6) would have automatically produced the time-dependent versions of the rod equations (see Chapter 7 and Problem 4). Finally, all the above reasoning may be extended to the situation in which the geometry remains that of the infinite slab, but the particles may have a discrete number of velocity states (see Problem 1).

In practice, where it is necessary ultimately to use computational methods in the resolution of transport problems, some sort of a discretization of (11.6) is necessary. This is basically equivalent to carrying out the kind of analysis described in this section. It is valuable, however, to do as much with the "continuous" version of the transport equation as possible, and we shall proceed with our discussion of (11.6).

4. BOUNDARY AND INITIAL CONDITIONS

Before going much further, it is necessary to investigate the side conditions that must accompany Eq. (11.6). The fundamental problem of transport theory as stated in the second paragraph of Section 2 of this chapter implies that the particle distribution in the medium is completely known at some initial time, $t=0$. Thus we may suppose that

$$N(\mathbf{r},\mathbf{v},0)=f(\mathbf{r},\mathbf{v}) \tag{11.11}$$

is specified throughout $X \times V$.

Moreover, we have stipulated that what becomes of a particle when it leaves the medium must be known. The simplest situation is that in which X is imbedded in a perfectly absorbing medium. In that event any particle which leaves the system immediately disappears from further consideration. The same sort of thing happens when X is a convex body surrounded by a vacuum. Here, again, there is no opportunity for the particle to reappear in X. (That the two types of boundary conditions can actually lead to different mathematical structures is by no means evident; it is rather a source of great surprise that different structures are encountered. For further information see [6–8]. We shall not be involved in such subtleties in the treatment given in this book.)

A much more complicated situation occurs when X is a nonconvex body embedded in a vacuum. In this case a particle, which has left the system, can obviously return and forms a part of the source on the surface of X.

Strictly speaking, then, $S(\mathbf{r},\mathbf{v},t)$ is not an extraneous source; it *does* depend upon the particle distribution existing in the medium. To avoid such complexities as much as possible, one ordinarily considers $S(\mathbf{r},\mathbf{v},t)$ as representing just the *internal* source distribution. Any source on the surface of X is given in the form of an appropriate boundary condition. Thus, if $\mathbf{r}_s$ is a point on the boundary of X and $\mathbf{v}_i$ is a velocity vector directed into the medium then we assume that $N(\mathbf{r}_s,\mathbf{v}_i,t)$ is given or obtainable. Our standard assumption of "good behavior" extends to the shape of regions so that such terms as $\mathbf{v}_i$ are meaningful.

In certain cases it will be desirable to consider media which extend to infinity. In such instances, conditions will be placed upon N so as to make the problem physically reasonable. It must be admitted, however, that conditions under which the transport equation has a unique solution are far from well understood.

In our actual applications of invariant imbedding the region X is almost always very simple in shape, and many of the difficulties to which we have alluded will not arise. In fact, so far as computational applications of imbedding to transport theory are concerned, only the simplest of media shapes can currently be handled. Yet there are enough problems of genuine practical interest here to make the technique interesting and valuable.

5. THE SPECIAL CASE OF SLAB GEOMETRY AND ONE SPEED

Equation (11.6) becomes much less formidable looking in the case of slab geometry provided we assume that all quantities involved depend on only one spatial variable, and the physical variables are time-independent. We take this spatial variable to be z. For further simplification let us assume that only one particle speed is possible. The resulting model, although a great idealization, is one of considerable physical interest and importance.

In Eq. (11.6) we suppress all dependence of σ, K, and S on x, y, and t. We also suppose that σ is independent of $\mathbf{v}$, and take $v=c$. (Here c is not necessarily meant to be the speed of light.) The particle density, N, may still be a function of time, t. It is rather apparent that the most troublesome quantity still remaining is the angular dependence hidden in $\mathbf{v} = (c\Omega_x, c\Omega_y, c\Omega_z)$. The standard direction cosines are not at all convenient, and it is much better to employ a new way of measuring direction. We therefore introduce polar and azimuthal angles, θ and φ. Consider the velocity vector $\mathbf{v}$ and denote the angle it makes with the positive z direction by θ. Hence $0 \leqslant \theta \leqslant \pi$. Now from any convenient reference plane measure the azimuthal angle φ so that $0 \leqslant \varphi < 2\pi$. (The clockwise or counterclockwise orientation is ordinarily not a matter of importance.) (See Figures 11.1 and 11.2.) It is customary to denote $\cos\theta$ by μ. Equation (11.6) may now be rewritten

$$\frac{1}{c}\frac{\partial N}{\partial t}(z,\mu,\varphi,t)+\mu\frac{\partial N}{\partial z}=-\sigma(z)N(z,\mu,\varphi,t)$$

$$+\sigma(z)\int_{-1}^{1}d\mu'\int_{0}^{2\pi}K(z,\mu',\varphi',\mu,\varphi)N(z,\mu',\varphi',t)\,d\varphi'+S(z,\mu,\varphi). \qquad (11.12)$$

Here we have taken advantage of the fact that the integral which occurs is really in ordinary spherical coordinates and $\sin\theta\, d\theta\, d\varphi=-d\mu\, d\varphi$, $-1\leqslant\mu\leqslant+1$.

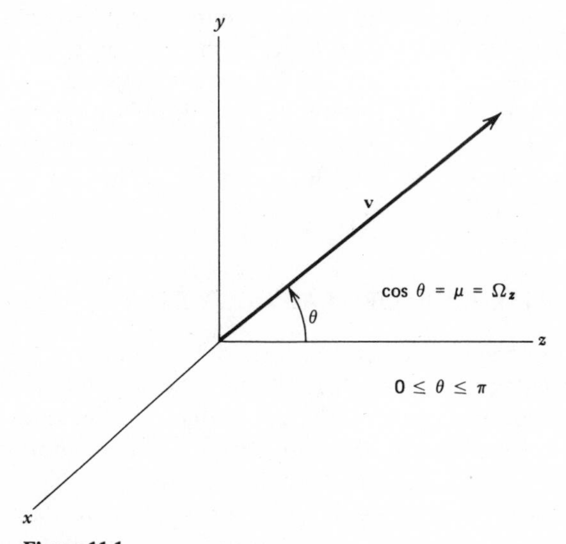

Figure 11.1.

In many cases of physical interest the angular dependence of K is much simpler than that indicated. Often K depends only upon the angle between $\mathbf{v}'$ and $\mathbf{v}$. (This is the case when the material of which the transport medium is made has no "sense of direction" of its own. Such was the assumption made in the study of the original rod problem of Chapter 1.) If we denote the angle between these two velocity vectors by $\bar{\theta}$ then the standard formulas of spherical trigonometry give

$$\cos\bar{\theta}=\mu\mu'+\sqrt{1-\mu^2}\,\sqrt{1-\mu'^2}\,\cos(\varphi-\varphi'). \qquad (11.13)$$

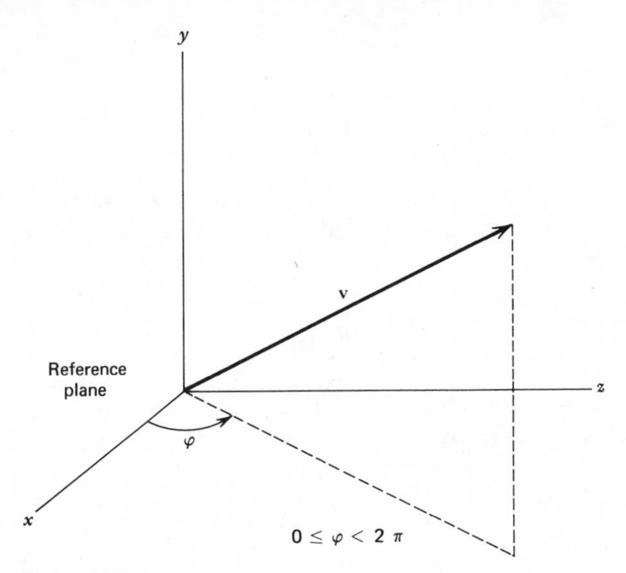

Figure 11.2.

We shall henceforth make the assumption that K depends only upon this angle and write, continuing to use our somewhat awkward notation,

$$K = K(z, \cos\bar{\theta}). \tag{11.14}$$

Frequently, boundary conditions are such that the dependence of N upon φ is of no great interest. Let us define

$$n(z,\mu,t) = \int_0^{2\pi} N(z,\mu,\varphi,t)\, d\varphi. \tag{11.15}$$

Because φ occurs in K only through the cosine term in Eq. (11.13) we can write

$$\int_0^{2\pi} K(z,\cos\bar{\theta})\, d\varphi = \tfrac{1}{2} k(z,\mu,\mu'). \tag{11.16}$$

Thus, integrating over φ yields

$$\frac{1}{c}\frac{\partial n}{\partial t} + \mu\frac{\partial n}{\partial z} = -\sigma(z) n(z,\mu,t)$$

$$+ \frac{\sigma(z)}{2}\int_{-1}^{1} k(z,\mu,\mu') n(z,\mu',t)\, d\mu' + s(z,\mu), \tag{11.17}$$

where

$$s(z,\mu)=\int_0^{2\pi} S(z,\mu,\varphi)\,d\varphi. \tag{11.18}$$

The simplest physical case of all is that in which k is completely independent of angle—the so-called *isotropic* case. In this situation it is customary to write

$$k=\frac{\gamma(z)}{4\pi}. \tag{11.19}$$

The quantity $\gamma(z)$ has physical significance. Suppose we integrate k over all θ and all φ—that is, over all directions *out* of a collision. Then

$$\int_0^{2\pi}\int_0^{\pi} k\sin\theta\,d\theta\,d\varphi=\int_0^{2\pi} d\varphi\int_{-1}^{1}\frac{\gamma(z)\,d\mu}{4\pi}=\gamma(z). \tag{11.20}$$

Thus, $\gamma(z)$ is actually the average total number of particles emerging from a collision. The transport equation now simplifies to

$$\frac{1}{c}\frac{\partial n}{\partial t}+\mu\frac{\partial n}{\partial z}=-\sigma(z)n(z,\mu,t)+\frac{\sigma(z)\gamma(z)}{2}\int_{-1}^{1} n(z,\mu',t)\,d\mu'+s(z,\mu). \tag{11.21}$$

As yet we have said little about side conditions. In the slab problem one often assumes that a particle which leaves the slab cannot return and that no particles enter the system at either face. Thus, if the surfaces of the slab are at $z=0$ and $z=a$, the conditions appropriate to (11.17) are

$$n(0,\mu,t)=0, \qquad 0<\mu\leqslant 1, \tag{11.22a}$$

$$n(a,\mu,t)=0, \qquad -1\leqslant\mu<0. \tag{11.22b}$$

6. THE TIME-INDEPENDENT SLAB PROBLEM VIA INVARIANT IMBEDDING—THE PERTURBATION APPROACH

In his original work Ambarzumian [9] applied his method of invariance to a time-independent problem having considerable similarity to the one we have been discussing. However, as an astrophysicist, he was interested in the case $\gamma \leqslant 1$. Moreover, since he was attempting to study the reflecting properties of planetary atmospheres, which are in general quite thick, he investigated a semi-infinite half-space geometry rather than a slab. His results were obtained essentially by purely physical reasoning—that is, particle-counting. This approach was also used in the analysis of the slab problem in [10]. It seems, therefore, a matter of some interest to apply one or more of the other approaches to invariant imbedding that we have devised in the preceding chapters to the study of the Ambarzumian model. In this section we shall use the perturbation technique outlined in Chapter 2, applying it first to the slab and then specializing the result to the semi-infinite half-space. In the following section the Riccati transformation method will be used to study a similar problem.

Let us assume a time-independent process so that n becomes a function of z and μ alone. As noted in Problem 4 of Chapter 1 we may remove the parameter σ completely by the so-called optical depth transformation. This is basically equivalent to taking $\sigma \equiv 1$. Strictly as a matter of convenience we shall assume that γ is a constant. Finally, at the risk of introducing some confusion, we shall suppose that the right side of the slab is at $z = x$. This is in agreement with the earlier chapters and should not really cause any discomfort since the x direction (or x coordinate) used in the present chapter plays no role in the problem under consideration. Finally, we assume that there are no *internal* sources, that no particles enter the slab at the left side, $z = 0$, and that the input at the right side is $Q(\mu)$ particles per square centimeter per second. The system (11.21) and (11.22) is therefore replaced by

$$\mu \frac{\partial n}{\partial z} = -n(z,\mu) + \frac{\gamma}{2}\int_{-1}^{1} n(z,\mu')\,d\mu', \tag{11.23a}$$

$$n(0,\mu) = 0, \qquad 0 < \mu \leqslant 1, \tag{11.23b}$$

$$n(x,\mu) = Q(\mu), \qquad -1 \leqslant \mu < 0. \tag{11.23c}$$

(As in Section 3 "density" and "flux" seem in the foregoing to be treated interchangeably. That there is no difficulty is illuminated by Problem 7.)

We notice at once that the familiar u and v type functions of previous work seem to be missing. To introduce their analogs we define

$$n^{+}(z,\mu) = n(z,\mu), \qquad 0 < \mu \leqslant 1, \tag{11.24a}$$

$$n^{-}(z,\mu) = n(z,\mu), \qquad -1 \leqslant \mu < 0, \tag{11.24b}$$

and rewrite (11.23a) as a pair of equations

$$\frac{\partial n^+}{\partial z} = -\frac{n^+(z,\mu)}{\mu} + \frac{\gamma}{2\mu}\int_{-1}^{0} n^-(z,\mu')\,d\mu' + \frac{\gamma}{2\mu}\int_{0}^{1} n^+(z,\mu')\,d\mu', \tag{11.25a}$$

$$-\frac{\partial n^-}{\partial z} = -\frac{n^-(z,\mu)}{|\mu|} + \frac{\gamma}{2|\mu|}\int_{-1}^{0} n^-(z,\mu')\,d\mu' + \frac{\gamma}{2|\mu|}\int_{0}^{1} n^+(z,\mu')\,d\mu'. \tag{11.25b}$$

The side conditions may now be rewritten also:

$$n^+(0,\mu)=0, \tag{11.25c}$$

$$n^-(x,\mu)=Q(\mu). \tag{11.25d}$$

This set now appears somewhat more familiar, especially when one recognizes that $n^+(z,\mu)$ represents the population of right-moving particles and is therefore analogous to u; similarly, $n^-(z,u)$ can be compared to v.

To implement the perturbation method we improve our notation somewhat, now writing

$$n^+ = n^+(z,\mu,x), \qquad n^- = n^-(z,\mu,x). \tag{11.26}$$

Finally, we must have an analog for the reflection function, which we shall call R. We attempt to define such a function by asking that

$$n^+(x,\mu,x) = \int_{-1}^{0} R(x,\mu,\mu')Q(\mu')\,d\mu'. \tag{11.27}$$

From a physical viewpoint, $R(x,\mu,\mu_0)$ is the reflection that would result if the input on the right surface of the slab were a delta function, $\delta(\mu'-\mu_0)$, $-1 \leqslant \mu_0 < 0$. The existence of such an R function is reasonably clear on physical grounds, provided γ and x are such that the slab cannot go critical. (Physically, this can never occur when $\gamma \leqslant 1$.) The mathematical proof of the existence of R is by no means obvious. It can be obtained by assuming the existence of such a function, determining the equations that it must satisfy, demonstrating that a unique solution to these equations exists, and then proving that the solution does have the requisite property given by (11.27). We shall not carry out the entire program though the reader will be lead through some of the details in the problems.

We begin by differentiating equations (11.25a, b) with respect to x, using the subscript notation for partial differentiation and making the assumption that $n_{13}^{\pm}(z,\mu,x)=n_{31}^{\pm}(z,\mu,x)$:

$$n_{31}^{+}=-\frac{n_3^{+}(z,\mu,x)}{\mu}+\frac{\gamma}{2\mu}\int_{-1}^{0}n_3^{-}(z,\mu',x)\,d\mu'$$

$$+\frac{\gamma}{2\mu}\int_{0}^{1}n_3^{+}(z,\mu',x)\,d\mu', \tag{11.28a}$$

$$-n_{31}^{-}=-\frac{n_3^{-}(z,\mu,x)}{|\mu|}+\frac{\gamma}{2|\mu|}\int_{-1}^{0}n_3^{-}(z,\mu',x)\,d\mu'$$

$$+\frac{\gamma}{2|\mu|}\int_{0}^{1}n_3^{+}(z,\mu',x)\,d\mu'. \tag{11.28b}$$

Next, we differentiate the conditions (11.25c, d), recalling that they really have the form $n^{+}(0,\mu,x)=0$, $n^{-}(x,\mu,x)=Q(\mu)$ and that Q is independent of x:

$$n_3^{+}(0,\mu,x)=0, \tag{11.28c}$$

$$n_3^{-}(x,\mu,x)=-n_1^{-}(x,\mu,x). \tag{11.28d}$$

Equations (11.28) are of precisely the same form as (11.25) when viewed as equations for the functions $n_3^{\pm}(z,\mu,x)$. From (11.27) then

$$n_3^{+}(x,\mu,x)=-\int_{-1}^{0}R(x,\mu,\mu')n_1^{-}(x,\mu',x)\,d\mu'. \tag{11.29}$$

If we differentiate (11.27) with respect to x we obtain

$$n_1^{+}(x,\mu,x)+n_3^{+}(x,\mu,x)=\int_{-1}^{0}R_1(x,\mu,\mu')Q(\mu')\,d\mu'. \tag{11.30}$$

Using (11.25), (11.27), (11.29), and (11.30) we shall next obtain a very complicated an unwieldy set of expressions which will eventually contain only the functions R and Q. Let us begin by evaluating (11.25a) at $z=x$:

$$n_1^{+}(x,\mu,x)=-\frac{n^{+}(x,\mu,x)}{\mu}+\frac{\gamma}{2\mu}\int_{-1}^{0}n^{-}(x,\mu',x)\,d\mu'+\frac{\gamma}{2\mu}\int_{0}^{1}n^{+}(x,\mu',x)\,d\mu'$$

$$=-\frac{1}{\mu}\int_{-1}^{0}R(x,\mu,\mu')Q(\mu')\,d\mu'+\frac{\gamma}{2\mu}\int_{-1}^{0}Q(\mu')\,d\mu'$$

$$+\frac{\gamma}{2\mu}\int_{0}^{1}d\mu'\int_{-1}^{0}R(x,\mu',\mu'')Q(\mu'')\,d\mu''. \tag{11.31}$$

Next, we turn to (11.29) and evaluate the integrand by setting $z=x$ in (11.25b)

$$n_3^+(x,\mu,x)=\int_{-1}^{0} R(x,\mu,\mu')\left\{-\frac{n^-(x,\mu',x)}{|\mu'|}+\frac{\gamma}{2|\mu'|}\int_{-1}^{0} n^-(x,\mu'',x)\,d\mu''\right.$$

$$\left.+\frac{\gamma}{2|\mu'|}\int_0^1 n^+(x,\mu'',x)\,d\mu''\right\}d\mu'$$

$$=\int_{-1}^{0} R(x,\mu,\mu')\,d\mu'\left\{-\frac{Q(\mu')}{|\mu'|}+\frac{\gamma}{2|\mu'|}\int_{-1}^{0} Q(\mu'')\,d\mu''\right.$$

$$\left.+\frac{\gamma}{2|\mu'|}\int_0^1 d\mu''\int_{-1}^{0} R(x,\mu'',\mu''')Q(\mu''')\,d\mu'''\right\}. \tag{11.32}$$

Now we may add (11.31) to (11.32) and use (11.30):

$$\int_{-1}^{0} R_1(x,\mu,\mu')Q(\mu')\,d\mu'=-\frac{1}{\mu}\int_{-1}^{0} R(x,\mu,\mu')Q(\mu')\,d\mu'$$

$$+\frac{\gamma}{2\mu}\int_{-1}^{0} Q(\mu')\,d\mu'+\frac{\gamma}{2\mu}\int_0^1 d\mu'\int_{-1}^{0} R(x,\mu',\mu'')Q(\mu'')\,d\mu''$$

$$-\int_{-1}^{0}\frac{R(x,\mu,\mu')}{|\mu'|}Q(\mu')\,d\mu'+\frac{\gamma}{2}\int_{-1}^{0} R(x,\mu,\mu')\frac{d\mu'}{|\mu'|}\int_{-1}^{0} Q(\mu'')\,d\mu''$$

$$+\frac{\gamma}{2}\int_{-1}^{0} R(x,\mu,\mu')\frac{d\mu'}{|\mu'|}\int_0^1 d\mu''\int_{-1}^{0} R(x,\mu'',\mu''')Q(\mu''')\,d\mu'''. \tag{11.33}$$

This is our desired equation. Taking care not to confuse integration variables, we write it in the form:

$$\int_{-1}^{0} G(x,\mu,\mu_0)Q(\mu_0)\,d\mu_0=0. \tag{11.34}$$

Now Eq. (11.34) is to hold for *all* input functions Q. We therefore require that

$$G(x,\mu,\mu_0)=0,\qquad 0<\mu\leqslant 1,\qquad -1\leqslant \mu_0<0. \tag{11.35}$$

In explicit form (11.35) becomes

$$\frac{\partial R}{\partial x}(x,\mu,\mu_0)=\frac{\gamma}{2\mu}-\left(\frac{1}{\mu}+\frac{1}{|\mu_0|}\right)R(x,\mu,\mu_0)$$

$$+\frac{\gamma}{2\mu}\int_0^1 R(x,\mu',\mu_0)\,d\mu'+\frac{\gamma}{2}\int_{-1}^0 R(x,\mu,\mu'')\frac{d\mu''}{|\mu''|}$$

$$+\frac{\gamma}{2}\int_{-1}^0 R(x,\mu,\mu'')\frac{d\mu''}{|\mu''|}\int_0^1 R(x,\mu',\mu_0)\,d\mu', \qquad (11.36)$$

a partial differential integro-equation for the function R. If $\sigma \neq 1$, then the left side of (11.36) becomes $(1/\sigma)(\partial R/\partial x)$; the right side does not change.

Obviously, side conditions of some sort are needed. As in the simple rod case we argue with a system of "size" zero—that is, we choose $x=0$. Clearly we must have $n^+(0,\mu,0)=0$, regardless of the input $Q(\mu)$. From (11.27) the condition

$$R(0,\mu,\mu_0)=0 \qquad (11.37)$$

follows readily.

It is not possible to state a priori any condition on the maximum value of x for which the system (11.36) and (11.37) is valid—that is a function of γ. For $\gamma \leqslant 1$ one expects a solution to exist for all x, and that is indeed the case. When $\gamma > 1$, which can occur in fissionable materials, one expects that the system will have a critical size, x_{cr}. It is reasonable to suppose that the system will then have a solution for all $x < x_{cr}$. We leave some of these matters to the problems.

Before proceding to the next section, let us pause long enough to obtain the original Ambarzumian result. The astrophysical model under study in his research involved $\gamma \leqslant 1$, and a semi-infinite half-space. In the work that we have just done the fact that the left side of the slab was at $z=0$ was used only in obtaining the condition (11.37). It had nothing to do with the derivation of (11.36). We may suppose then that the left side of the slab has receded to $z=-\infty$. But then the exact position of the right face is of no consequence. That is, R becomes independent of x. If we write in this situation

$$R=\tilde{R}(\mu,\mu_0), \qquad (11.38)$$

then (11.36) reduces to

$$\frac{1}{\gamma}\left(\frac{1}{\mu}+\frac{1}{|\mu_0|}\right)\tilde{R}(\mu,\mu_0)=\frac{1}{2\mu}+\frac{1}{2\mu}\int_0^1\tilde{R}(\mu',\mu_0)\,d\mu'$$

$$+\frac{1}{2}\int_{-1}^{0}\tilde{R}(\mu,\mu'')\frac{d\mu''}{|\mu''|}$$

$$+\frac{1}{2}\int_{-1}^{0}\tilde{R}(\mu,\mu'')\frac{d\mu''}{|\mu''|}\int_0^1\tilde{R}(\mu',\mu_0)\,d\mu'. \qquad (11.39)$$

We shall have a bit more to say about this result later in this chapter. As remarked earlier, it was originally obtained by particle-counting type arguments. Ambarzumian noted that when a slab of homogeneous material is added to a semi-infinite half-space of the same material one still has a semi-infinite half-space of homogeneous material. Hence the new system must have exactly the same reflection properties as the original. This explains the origins of the term "invariance" commonly used by Ambarzumian, Chandasekhar, and their followers.

7. THE TIME-INDEPENDENT SLAB PROBLEM VIA INVARIANT IMBEDDING—THE RICCATI TRANSFORMATION

In this section we shall use the Riccati method (See Chapter 2) to study the slab problem. For variety, let us suppose that internal sources do exist in the system and that in addition to the input $Q(\mu)$ at $z=x$ there is an input $W(\mu)$ at $z=0$, $0<\mu\leqslant 1$. We continue to assume the isotropic case with γ still a constant. As noted earlier, there is no loss of generality in assuming $\sigma=1$. (We refrain from succumbing to the temptation at greater generality that confronts us here: namely, to include an arbitrary kernel $k(z,\mu,\mu')$ in the problem formulation. To do so is to considerably increase the amount of algebra with which one must contend. As is our custom, we consign such complexities to the problem set.)

Recall that in the Riccati transformation method, explicit use of the parameter x is not necessary, at least not until rather late in the analysis. However, it is still desirable to decompose the source term into two parts:

$$s^+(z,\mu)=s(z,\mu), \qquad 0<\mu\leqslant 1,$$
$$s^-(z,\mu)=s(z,\mu), \qquad -1\leqslant\mu<0. \qquad (11.40)$$

Equations (11.25) are now replaced by

$$\frac{\partial n^+}{\partial z} = -\frac{n^+(z,\mu)}{\mu} + \frac{\gamma}{2\mu}\int_{-1}^{0} n^-(z,\mu')\,d\mu' + \frac{\gamma}{2\mu}\int_{0}^{1} n^+(z,\mu')\,d\mu' + \frac{s^+(z,\mu)}{\mu}, \tag{11.41a}$$

$$-\frac{\partial n^-}{\partial z} = -\frac{n^-(z,\mu)}{|\mu|} + \frac{\gamma}{2|\mu|}\int_{-1}^{0} n^-(z,\mu')\,d\mu' + \frac{\gamma}{2|\mu|}\int_{0}^{1} n^+(z,\mu')\,d\mu' + \frac{s^-(z,\mu)}{|\mu|}, \tag{11.41b}$$

$$n^+(0,\mu) = W(\mu), \tag{11.41c}$$

$$n^-(x,\mu) = Q(\mu). \tag{11.41d}$$

We now seek an analog of (2.40). The success obtained in the previous section in using (11.27) suggests that we try

$$n^+(z,\mu) = \int_{-1}^{0} \rho(z,\mu,\mu')n^-(z,\mu')\,d\mu' + w(z,\mu). \tag{11.42}$$

(For a rather complete discussion of the physical significance of this transformation see [11]. The matter is also elucidated in the problem section.) We differentiate (11.42) with respect to z and use (11.41a, b):

$$\frac{-n^+(z,\mu)}{\mu} + \frac{\gamma}{2\mu}\int_{-1}^{0} n^-(z,\mu')\,d\mu' + \frac{\gamma}{2\mu}\int_{0}^{1} n^+(z,\mu')\,d\mu' + \frac{s^+(z,\mu)}{\mu}$$
$$= \int_{-1}^{0} \rho_1(z,\mu,\mu')n^-(z,\mu')\,d\mu' + \int_{-1}^{0} \rho(z,\mu,\mu')\,d\mu'$$
$$\times\left\{\frac{n^-(z,\mu')}{|\mu'|} - \frac{\gamma}{2|\mu'|}\int_{-1}^{0} n^-(z,\mu'')\,d\mu'' - \frac{\gamma}{2|\mu'|}\int_{0}^{1} n^+(z,\mu'')\,d\mu'' - \frac{s^-(z,\mu')}{|\mu'|}\right\} + w_1(z,\mu). \tag{11.43}$$

In (11.43) we replace n^+ wherever it occurs by the expression (11.42). After a considerable amount of tedious manipulation we obtain:

$$\int_{-1}^{0}\left[-\frac{\rho(z,\mu,\mu_0)}{\mu}+\frac{\gamma}{2\mu}+\frac{\gamma}{2\mu}\int_0^1\rho(z,\mu',\mu_0)\,d\mu'\right.$$

$$-\rho_1(z,\mu,\mu_0)-\frac{\rho(z,\mu,\mu_0)}{|\mu_0|}+\frac{\gamma}{2}\int_{-1}^{0}\rho(z,\mu,\mu'')\frac{d\mu''}{|\mu''|}$$

$$\left.+\frac{\gamma}{2}\int_{-1}^{0}\rho(z,\mu,\mu'')\frac{d\mu''}{|\mu''|}\int_0^1\rho(z,\mu',\mu_0)\,d\mu'\right]n^-(z,\mu_0)\,d\mu_0$$

$$=+\frac{w(z,\mu)}{\mu}-\frac{s^+(z,\mu)}{\mu}-\frac{\gamma}{2\mu}\int_0^1 w(z,\mu')\,d\mu'$$

$$-\frac{\gamma}{2}\int_{-1}^{0}\rho(z,\mu,\mu'')\frac{d\mu''}{|\mu''|}\int_0^1 w(z,\mu')\,d\mu'$$

$$-\int_{-1}^{0}\rho(z,\mu,\mu'')\frac{s^-(z,\mu'')}{|\mu''|}\,d\mu''+w_1(z,\mu). \qquad (11.44)$$

Recall that ρ and w are still quite arbitrary. We choose ρ so as to make the left side of (11.44) vanish identically:

$$\rho_1(z,\mu,\mu_0)=\frac{\gamma}{2}-\left(\frac{1}{\mu}+\frac{1}{|\mu_0|}\right)\rho(z,\mu,\mu_0)$$

$$+\frac{\gamma}{2\mu}\int_0^1\rho(z,\mu',\mu_0)\,d\mu'+\frac{\gamma}{2}\int_{-1}^{0}\rho(z,\mu,\mu'')\frac{d\mu''}{|\mu''|}$$

$$+\frac{\gamma}{2}\int_{-1}^{0}\rho(z,\mu,\mu'')\frac{d\mu''}{|\mu''|}\int_0^1\rho(z,\mu',\mu_0)\,d\mu'. \qquad (11.45)$$

This, of course, yields

$$w_1(z,\mu)=-\frac{w(z,\mu)}{\mu}+\frac{\gamma}{2}\int_{-1}^{0}\rho(z,\mu,\mu'')\frac{d\mu''}{|\mu''|}\int_0^1 w(z,\mu')\,d\mu'$$

$$+\frac{s^+(z,\mu)}{\mu}+\int_{-1}^{0}\rho(z,\mu,\mu'')\frac{s^-(z,\mu'')}{|\mu''|}\,d\mu''. \qquad (11.46)$$

Notice that Eq. (11.45) is identical to (11.36). This suggests that we arbitrarily impose on ρ the condition

$$\rho(0,\mu,\mu_0)=0 \tag{11.47}$$

thus obtaining $\rho(z,\mu,\mu_0)\equiv R(z,\mu,\mu_0)$. Of course, to make the choice (11.47) we must be certain that the remaining conditions of the problem (11.41) can be satisfied. Using (11.42) and (11.47) we obtain

$$n^+(0,\mu)=w(0,\mu)=W(\mu). \tag{11.48}$$

Since (11.46) is a linear partial differential integro-equation in $w(z,\mu)$ the requirement (11,48) is a reasonable one. Only (11.41d) needs still to be dealt with. However, we notice that as yet the quantity x has not entered the picture. All equations derived by the Riccati method are equations in z. We now suppose that it is possible to integrate the equations for $\rho(z,\mu,\mu_0)$ and for $w(z,\mu)$ over the interval $0\leqslant z\leqslant x$. If this can be accomplished then we may rewrite (11.41b) with the use of (11.42) as

$$\begin{aligned} -\frac{\partial n^-}{\partial z} &= -\frac{n^-(z,\mu)}{|\mu|}+\frac{\gamma}{2|\mu|}\int_{-1}^{0} n^-(z,\mu')\,d\mu' \\ &\quad+\frac{\gamma}{2|\mu|}\int_0^1 d\mu'\left[\int_{-1}^{0} R(z,\mu',\mu'')n^-(z,\mu'')\,d\mu''+w(z,\mu')\right] \\ &\quad+\frac{s^-(z,\mu)}{|\mu|}, \end{aligned} \tag{11.49}$$

and integrate this equation *backwards* from $z=x$ to $z=0$ subject to the condition (11.41d). Finally, $n^+(z,\mu)$ may be found for any z, $0\leqslant z\leqslant x$, and any μ, $0<\mu\leqslant 1$, by using Eq. (11.42). This resolves the problem (11.41) completely.

It cannot be denied that the above scheme contains a number of mathematical "holes." We have had to assume the solubility of several partial differential integro-equations. However, only one of these, the equation for ρ or, equivalently, for R, is nonlinear. The analysis of the other equations is not difficult provided s^+ and s^- are reasonably well-behaved functions.

It is important to again note that the Riccati method forces us to the observation that R is really a meaningful function internal to the system as well as at its surface. This was known rather early in the study of the invariant imbedding method. The equations of Chapter 3 represent the ultimate (at least for the present!) culmination of the fact that the reflection and transmission functions are well defined and useful at all internal points of the system under study.

We note in concluding this section that if both $s(z,\mu)=0$ and $W(\mu)\equiv 0$, then $w(z,\mu)\equiv 0$ is a solution of (11.46)–(11.48). In this case (11.42) reduces at $z=x$ to

$$n^+(x,\mu)=\int_{-1}^{0}\rho(x,\mu,\mu')n^-(x,\mu')\,d\mu'=\int_{-1}^{0}R(x,\mu,\mu')Q(\mu')\,d\mu', \quad (11.50)$$

in complete agreement with (11.27). If, however, $s(z,\mu)\equiv 0$, but $W(\mu)\not\equiv 0$ then $w(z,\mu)=0$ does not solve (11.46). The solution to the problem is still given, of course, by the overall scheme we have outlined for the solution of (11.41).

The above remarks point out again the fundamental physical importance of the reflection function for a transport system. Once $R(z,\mu,\mu')$ is known the rest of the quantities of physical interest may be found by solving (relatively) simple linear initial value problems. It seems quite possible that this fact could be used advantageously when doing actual experiments with transport phenomena.

8. A RETURN TO THE CASE OF THE SEMI-INFINITE HALF-SPACE

Although it is something of a digression from our main purposes, it would be almost an act of historical heresy not to investigate further the case of original interest to Ambarzumian and Chandrasekhar. It was the study of Eq. (11.39), and certain variants of it, that first called attention to the imbedding method and especially to the tremendous numerical advantages the method sometimes possesses.

Ambarzumian, and others before him, were interested in the reflection of a plane parallel beam of radiation (sunlight) from the foggy atmosphere of a planet like Venus. The obvious way to approach this problem is to integrate (11.23). Clearly, no analytical solution is likely to be found, so numerical integration is called for. But only in the last decade or so has it been possible to handle such an equation numerically, especially for x large, the physically reasonable case. Even after the reduction of the problem to the form (11.39), the numerical difficulties presented were still much more than the primitive calculators of the pre-World War II days could handle. The manipulations we are about to discuss further reduced the problem to one whose computation was entirely feasible on a desk calculator. Indeed, many of the very significant results obtained by Chandrasekhar were obtained with precisely such crude equipment. Thus the method of invariance reduced a totally intractable astrophysical problem to one that could be completely analyzed and computed.

It is convenient to put Eq. (11.39) into a somewhat different form. First, we set $\mu_0 = -\mu_1$, so that $0 < \mu_1 \leqslant 1$, and hence all integrations are taken over the interval $0 < \mu, \mu_1 \leqslant 1$. Next, we define $\tilde{S}(\mu, \mu_1)$ by

$$\tilde{R}(\mu, \mu_0) = \frac{\tilde{S}(\mu, \mu_1)}{2\mu}. \tag{11.51}$$

After some algebra this reduces (11.39) to

$$\begin{aligned}\frac{1}{\gamma}\left(\frac{1}{\mu} + \frac{1}{\mu_1}\right)\tilde{S}(\mu, \mu_1) &= 1 + \frac{1}{2}\int_0^1 \tilde{S}(\mu', \mu_1)\frac{d\mu'}{\mu'} \\ &\quad + \frac{1}{2}\int_0^1 \tilde{S}(\mu, \mu'')\frac{d\mu''}{\mu''} + \frac{1}{4}\int_0^1 \frac{\tilde{S}(\mu', \mu_1)\,d\mu'}{\mu'}\int_0^1 \tilde{S}(\mu, \mu'')\frac{d\mu''}{\mu''} \\ &= \left\{1 + \frac{1}{2}\int_0^1 \tilde{S}(\mu', \mu_1)\frac{d\mu'}{\mu'}\right\}\left\{1 + \frac{1}{2}\int_0^1 \tilde{S}(\mu, \mu'')\frac{d\mu''}{\mu''}\right\}. \end{aligned} \tag{11.52}$$

The structure of (11.52) suggests that we seek a solution of the form

$$\frac{1}{\gamma}\left(\frac{1}{\mu} + \frac{1}{\mu_1}\right)\tilde{S}(\mu, \mu_1) = H(\mu)H(\mu_1). \tag{11.53}$$

Substitution into (11.52) produces

$$\begin{aligned} H(\mu)H(\mu_1) &= \left\{1 + \frac{\gamma}{2}\mu_1 H(\mu_1)\int_0^1 \frac{H(\mu')\,d\mu'}{\mu' + \mu_1}\right\} \\ &\quad \times \left\{1 + \frac{\gamma}{2}\mu H(\mu)\int_0^1 \frac{H(\mu'')\,d\mu''}{\mu'' + \mu}\right\}, \end{aligned} \tag{11.54}$$

and setting $\mu = \mu_1$ gives

$$H^2(\mu) = \left\{1 + \frac{\gamma}{2}\mu H(\mu)\int_0^1 \frac{H(\mu')}{\mu + \mu'}\,d\mu'\right\}^2. \tag{11.55}$$

If we now take the (positive) square root in (11.55) we obtain

$$\frac{1}{H(\mu)} = 1 - \frac{\gamma}{2}\mu\int_0^1 \frac{H(\mu')}{\mu + \mu'}\,d\mu'. \tag{11.56}$$

This is the famous nonlinear integral equation that was treated so effectively by Chandrasekhar. His first approach was to proceed by the method of quadrature, thus obtaining a set of nonlinear algebraic equations. Using intricate manipulations and great insight he obtained an analytic solution to the approximate problem, and employed this result to make the many calculations earlier referred to. He also obtained a semi-rigorous solution to the continuous version as posed in equation (11.56). Further work to put this analysis on a firm mathematical foundation has been done by a variety of researchers, many of whom looked at the somewhat more complicated problem involving various forms of anisotropy [3,4].

It should also be noted that a variety of interesting mathematical questions have been avoided in our brief treatment. Certainly the "separation of variables" method used in (11.53), while ultimately successful, may not give the only physically acceptable solution to the fundamental problem (11.52). We have chosen the positive square root of equation (11.55); this is not hard to justify. However, we have also omitted any discussion of the fact that (11.56) may not have a solution, or that it may have more than one solution, even when one restricts himself to a reasonable class of functions. This is indeed a problem, for it turns out that only for $\gamma=1$ is the solution to Eq. (11.56) unique. (Recall that the physics of the problem demands $\gamma \leqslant 1$.) The extraneous solution that arises when $\gamma<1$ can be discarded on the grounds that it implies that more particles emerge from the half space than are injected, a physical impossibility when $\gamma \leqslant 1$. The situation is much more complex in more realistic cases and has been studied extensively by T. Mullikin [12]. An excellent treatment of many of the mathematical questions that arise in the half-space transport problem is given in the book by I. W. Busbridge [3].

9. INVARIANT IMBEDDING AS A CALCULATIONAL DEVICE FOR TRANSPORT PROBLEMS IN A SLAB

As indicated in the preface to this book, it is not our intention to dwell at great length on the computational aspects of the imbedding method. However, a few words about such matters are appropriate here. We have pointed out in the foregoing section that many astrophysical problems of great interest were resolved by the application of the method to the half-space geometry. Such a geometry becomes of little interest in the case $\gamma>1$, nor is it of great value when one is attempting to study shielding problems in neutron physics in which the thickness of the shield is a matter of very definite importance. In such instances one must turn to equations like (11.36) and (11.46), or, more generally, to their extensions involving anisotropic interactions, energy dependence, and the like. (See Problems 9 and 13.)

It is certainly valid to ask if the invariant imbedding formulation provides any advantages over the classical Boltzmann equation. The answer is to a certain extent based on actual computational experience, although it is possible to investigate some relatively simple problems analytically (see Problem 17). In fact, some researchers report that the imbedding method has produced computational answers in certain problems that would not yield at all to the more classical approach even with present day computers. It is clear from the form of the imbedding equations that even when the standard methods of transport theory give perfectly satisfactory numerical answers, the invariant imbedding approach may be advantageous. For example, if one is primarily interested in the reflection properties of a medium that varies in thickness, then the R equation immediately supplies this information. The Boltzmann technique requires the separate solution of each problem in the family, and the subsequent evaluation of the density function at the surface of the slab.

Some rather extensive sets of data computed by the invariant imbedding method exist for certain transport problems. We refer the interested reader to the references [13–15]. It must also be pointed out that a great deal of such computation has been done without finding its way into the standard public literature and is hidden away in technical reports (some classified).

10. TRANSPORT THEORY IN OTHER GEOMETRIES

Thus far we have confined our discussion in this chapter wholly to the slab geometry, at least so far as the imbedding scheme is concerned. The fundamental Boltzmann equation is valid in any geometry (provided the medium itself is not changing or moving). The vector form given in Eq. (11.6) is most appropriate for treating more general configurations.

A wide variety of attempts has been made to write down invariant imbedding formulations of transport problems in nonslab configurations. A large number of the equations obtained are simply incorrect. Errors are especially common in those results obtained by particle-counting. [This remark, unfortunately, pertains to the imbedding equations for transport in the sphere and cylinder geometries found in the book written by one of the present authors [2] (see also [16]).] Some rather extensive calculations have been done using these faulty formulations. Basically, both the sphere and the infinite cylinder should yield to the more analytical methods that we have presented. There are, however, rather subtle difficulties involved in assigning side conditions. Moreover, there are usually analytical and computational problems at the center of the sphere and on the axis of the cylinder. Because this whole problem area is not reasonably well resolved at the present time, we prefer to leave the matter open. It does seem almost certain that sufficiently careful analysis of the sphere and cylinder will yield correct and reasonably tractable invariant imbedding equations. It also seems just as certain that their computation via finite difference methods will be a considerable challenge. However, it must be pointed out that these geometries are also not at all easy to study numerically in the classical Boltzmann formulation, and many of the present calculational schemes leave much to be desired. (Some, for example, fail to preserve total numbers of particles, even in the physically conservative case.)

More complicated geometries may be almost inaccessible by the strict imbedding method—at least so far as actual numerics are concerned. Equations may be written down for rather complicated configurations, but they are quite beyond the capabilities of the present-day computer. Some effort has been made at studying transport in parallelepipeds and the like by approximating the behavior in some directions by classical means—such as diffusion theory—and then employing the imbedding technique in just the one remaining spatial direction [14]. It has also been pointed out that transform methods may sometimes be used to remove dependence on certain spatial coordinates, leaving a problem amenable to the invariant imbedding approach. The difficulty here, of course, is that eventually the inverse transform of the result will have to be taken. This is usually difficult numerically.

11. TIME-DEPENDENT TRANSPORT IN A SLAB GEOMETRY

We turn now to the problem of time-dependent transport. In view of the comments made in the preceding section it is not at all surprising that we confine our attention to the relatively simple slab geometry. Even in this case the invariant imbedding equations for the time-dependent case were very elusive until the transform method described in Chapter 7 was developed. We shall use that device in our present treatment, reducing the problem to one that essentially fits the analyses of the previous sections, and then performing the Laplace inversion. The resulting equations are of an interesting and unusual structure. As noted later, their direct computational value is probably not great (see, however, [17]).

Again it might be argued that since, for numerical purposes, discretization of the problem at least so far as angular dependence is concerned is absolutely essential, there is little point in carrying thru the details of the argument to be given below. One can just use the matrix approach indicated in Chapter 7. We feel, however, that our results are of value in themselves and that at some future time might actually be useful in obtaining information about realistic time-dependent transport models.

The basic form of the Boltzmann equation that we shall study is given by (11.21). Recall that this governs the isotropic case. Our discussion could easily be extended to the anisotropic case provided the kernel k remains time-independent. However, as in the analysis done in the earlier part of this chapter, the details become very messy and the whole matter is best left to the Problem section (see Problem 18). For the sake of simplicity we suppose $\sigma=1$ and $\gamma=$constant. We shall also assume that no internal sources are present so that $s(z,\mu)$, which *could* actually be $s(z,\mu,t)$, will be taken as zero. We shall assume an input of a "parallel beam" of particles on the right face of the slab, $z=x$; that is, all impinging particles will be assumed to have the direction μ_0, $-1\leqslant\mu_0<0$. Moreover, we suppose that this beam is merely an impulse, falling on the slab face instantaneously at time zero. We so normalize this input that the boundary condition at $z=x$ takes the form:

$$n^-(x,\mu,t)=\delta(t)\delta(\mu-\mu_0). \tag{11.57}$$

[Our notation will parallel that of previous sections and we shall not explain the precise meaning of each new function introduced unless the analogy is not completely transparent. We note also that the normalization implicit in (11.57) is again a matter of convenience. We wish to emerge from our calculations with equations as nearly like those of Section 6 as possible. Different normalizations, inputs, and the like, can easily be used (see Problems 19).]

As in the case of the rod (see Chapter 7) it is physically sensible—and perhaps intuitively most easily understood—if we concentrate on the *total* number of particles emergent from the right face of the slab in a given direction from time zero to time T. We therefore define

$$cN^{\pm}(z,\mu,T)=c\int_0^T n^{\pm}(z,\mu,t)\,dt. \tag{11.58}$$

[Observe that this is not the N of (11.1).] Assuming no particles in the slab at time zero, none entering the slab from the left side at any time, and none entering on the right at any time greater than zero we readily find by using (11.21), the general methods of Section 6 and Eq. (11.57), that we must examine the following problem:

$$\frac{1}{c}\frac{\partial N^+}{\partial T}+\mu\frac{\partial N^+}{\partial z}=-N^+(z,\mu,T)+\frac{\gamma}{2}\int_{-1}^{0}N^-(z,\mu',T)\,d\mu' +\frac{\gamma}{2}\int_0^1 N^+(z,\mu',T)\,d\mu',$$

$$\frac{1}{c}\frac{\partial N^-}{\partial T}-|\mu|\frac{\partial N^-}{\partial z}=-N^-(z,\mu,T)+\frac{\gamma}{2}\int_{-1}^{0}N^-(z,\mu',T)\,d\mu' +\frac{\gamma}{2}\int_0^1 N^+(z,\mu',T)\,d\mu', \tag{11.59}$$

$$N^+(0,\mu,T)=0,\qquad T\geqslant 0, \tag{11.59a}$$

$$N^-(x,\mu,T)=\delta(\mu-\mu_0),\qquad T>0, \tag{11.59b}$$

$$N^+(z,\mu,0)=N^-(z,\mu,0)=0,\qquad 0\leqslant z<x. \tag{11.59c}$$

The Laplace transform with respect to T is obviously called for:

$$L_T[N^{\pm}(z,\mu,T)]=\tilde{N}^{\pm}(z,\mu,s),$$

$$\frac{\partial\tilde{N}^+}{\partial z}=\frac{-[1+(s/c)]\tilde{N}^+}{\mu}+\frac{\gamma}{2\mu}\int_{-1}^{0}\tilde{N}^-(z,\mu',s)\,d\mu' +\frac{\gamma}{2\mu}\int_0^1\tilde{N}^+(z,\mu',s)\,d\mu',$$

$$-\frac{\partial\tilde{N}^-}{\partial z}=\frac{-[1+(s/c)]}{|\mu|}\tilde{N}^-+\frac{\gamma}{2|\mu|}\int_{-1}^{0}\tilde{N}^-(z,\mu',s)\,d\mu' +\frac{\gamma}{2|\mu|}\int_0^1\tilde{N}^+(z,\mu',s)\,d\mu', \tag{11.60}$$

$$\tilde{N}^+(0,\mu,s)=0, \tag{11.60a}$$

$$\tilde{N}^-(x,\mu,s)=\frac{1}{s}\delta(\mu-\mu_0). \tag{11.60b}$$

Equations (11.60) are in precisely the form obtainable from (11.25) after restoration of σ, provided we make the following identifications:

$$\sigma\to 1+\frac{s}{c},\qquad \gamma\to\frac{\gamma}{1+\frac{s}{c}},\qquad Q(\mu)\to\frac{\delta(\mu-\mu_0)}{s}. \tag{11.61}$$

Thus, associated with Eqs. (11.60) is a reflection equation that may be obtained directly from (11.36) upon making the identifications (11.61). We must remember that the left-hand side of (11.36) must be put in the form $(1/\sigma)(\partial R/\partial x)$ (as noted in the remark just after that equation) since now the σ-like function is definitely not unity. We shall also use the notation $R(x,\mu,\mu_0,s)$ to emphasize that the parameter s is of considerable importance. Hence,

$$\frac{\partial R}{\partial x}(x,\mu,\mu_0,s)=\frac{\gamma}{2\mu}-\left(1+\frac{s}{c}\right)\left(\frac{1}{\mu}+\frac{1}{|\mu_0|}\right)R$$
$$+\frac{\gamma}{2\mu}\int_0^1 R(x,\mu',\mu_0,s)\,d\mu'+\frac{\gamma}{2}\int_{-1}^{0} R(x,\mu,\mu'',s)\frac{d\mu''}{|\mu''|}$$
$$+\frac{\gamma}{2}\int_{-1}^{0} R(x,\mu,\mu'',s)\frac{d\mu''}{|\mu''|}\int_0^1 R(x,\mu',\mu_0,s)\,d\mu'. \quad (11.62)$$

Finally, we recall that the quantity of physical interest that we are now seeking is not simply $R(x,\mu,\mu_0,s)$ but rather $cN^+(x,\mu,T)$. Using appropriate changes in notation we note from (11.27) that

$$c\tilde{N}^+(x,\mu,s)=c\int_{-1}^{0} R(x,\mu,\mu',s)\frac{\delta(\mu'-\mu_0)}{s}\,d\mu'$$
$$=\frac{c}{s}R(x,\mu,\mu_0,s). \quad (11.63)$$

The equation satisfied by $\tilde{N}^+(x,\mu,s)\equiv\tilde{N}^+(x,\mu,\mu_0,s)$ is, thus,

$$\frac{\partial \tilde{N}^+}{\partial x}(x,\mu,\mu_0,s)=\frac{\gamma}{2\mu s}-\left(1+\frac{s}{c}\right)\left(\frac{1}{\mu}+\frac{1}{|\mu_0|}\right)\tilde{N}^+$$
$$+\frac{\gamma}{2\mu}\int_0^1 \tilde{N}^+(x,\mu',\mu_0,s)\,d\mu'+\frac{\gamma}{2}\int_{-1}^{0} \tilde{N}^+(x,\mu,\mu'',s)\frac{d\mu''}{|\mu''|}$$
$$+\frac{\gamma}{2}\int_{-1}^{0} s\tilde{N}^+(x,\mu,\mu'',s)\frac{d\mu''}{|\mu''|}\int_0^1 \tilde{N}^+(x,\mu',\mu_0,s)\,d\mu'. \quad (11.64)$$

It is now necessary to take the Laplace inverse of (11.64):

$$\frac{\partial N^+}{\partial x}+\frac{1}{c}\left(\frac{1}{\mu}+\frac{1}{|\mu_0|}\right)\frac{\partial N^+}{\partial T}+\left(\frac{1}{\mu}+\frac{1}{|\mu_0|}\right)N^+(x,\mu,\mu_0,T)$$

$$=\frac{\gamma}{2\mu}+\frac{\gamma}{2\mu}\int_0^1 N^+(x,\mu',\mu_0,T)\,d\mu'$$

$$+\frac{\gamma}{2}\int_0^T dt\int_{-1}^0 n^+(x,\mu,\mu'',t)\frac{d\mu''}{|\mu''|}\int_0^1 N^+(x,\mu',\mu_0,T-t)\,d\mu'. \quad (11.65)$$

Here we have introduced the rather obvious notation

$$N^+(x,\mu,T)=N^+(x,\mu,\mu_0,T) \quad \text{and} \quad n^+(x,\mu,T)=n^+(x,\mu,\mu_0,T). \quad (11.66)$$

The solution of (11.65) is subject to the conditions

$$N^+(0,\mu,\mu_0,T)=0, \qquad T\geqslant 0, \quad (11.65a)$$

$$N^+(x,\mu,\mu_0,0)=0, \qquad x\geqslant 0. \quad (11.65b)$$

It should be noted that the choice of boundary conditions and initial conditions again allows the use of Duhamel's principle to study more complicated situations. However, it is also appropriate to emphasize once more that the numerical solution of Eqs. (11.65) would be extremely difficult (the authors are not aware of any successful attempts at such calculations). It is much more reasonable to start from (11.62), replace the integrations by quadratures and then invert the Laplace transform numerically. (See [17].)

12. SUMMARY

In this chapter we have taken up the classical problem of transport theory or radiative transfer, the area in which invariant imbedding first scored notable successes. Since this is a field of physical science unfamiliar to many people—including some physicists—it has seemed necessary to dwell at some length on the background of the problems and the derivations of the classical equations before even beginning the imbedding treatment. We have found that the method is very well adapted to slab geometries and semi-infinite half-spaces, and have pointed out without going into much detail the difficulties with more complicated configurations. In many slablike problems the invariant imbedding approach has very definite numerical advantages over the classical devices and it is not unreasonable to anticipate success in more complicated situations as the method is better understood and computer technology improves.

PROBLEMS

1. Generalize the ideas of Section 3 to the case in which the particles are all of the same speed but can move in a finite number of directions. Do this by use of appropriate delta functions. Consider also the case in which the particles can move in only two directions (right and left) but can have a finite number of different energies.

Now combine the results obtained above to arrive at a version of the transport equation discretized in both energy and direction. Compare the answers with the many-state case studied in Chapter 1. Also, try to obtain similar sets of equations by replacing the integral term in the Boltzmann equation with a suitable quadrature approximation. Discuss possible computational uses of this approach.

2. Try to eliminate the use of the delta function in Section 3 and Problem 1 by introducing either the Stieltjes integral or the theory of distributions.

3. Describe physically the similarity between the "two-flow" slab analysis of Section 3 and the results for the rod obtained in Chapter 1. Observe that this reasoning really makes the rod model much more meaningful than may have seemed to be the case earlier.

4. Obtain the time-dependent rod model from the general Boltzmann equation by appropriate discretization. Generalize to many angles and many energies.

5. Consider the time-independent rod model but let the right end of the rod be at plus infinity with input at the left end. For f and b constants confirm that the problem is soluble if $f+b\leqslant 1$, insoluble if $f+b>1$. What would you expect to happen if f and b were functions of z but $f(z)+b(z)>1$ for all z? Can you prove this? Suppose $f(z)+b(z)>1$ only for some set of z values, with $f(z)+b(z)\leqslant 1$ off the set. Can you make any guesses as to the solubility of the problem?

Try to generalize, at least for f_{ij} and b_{ij} independent of z, to the n-state problem. (See Chapter 4.)

6. In Section 5, after certain assumptions were made about K, it was asserted that the φ dependence of N is of no great interest. Consider a plane-parallel beam of particles impinging at $z=x$ in the specific direction (θ_0,φ_0). Make a physical argument to show that after each particle makes one collision the dependence on φ_0 is no longer observable. Show mathematically how those particles that have made no collisions can be isolated from the others and treated separately. Note that this implies that in dealing with reflection functions for such models the angle φ_0 is unimportant, since a particle must suffer at least one collision to be reflected. What can you say about the corresponding transmission functions?

7. Resolve the matter of the apparent interchangeability of "flux" and "density" (see Section 6).

8. Obtain Eq. (11.36) from first principles using particle-counting. (See R. Bellman, R. E. Kalaba, and M. C. Prestrud, *Invariant Imbedding and Radiative Transfer in Slabs of Finite Thickness*, American Elsevier, New York, 1963.)

9. Obtain the analog of (11.36) when the assumption of isotropy is dropped.

10. The structure of Eq. (11.42) is not surprising in the light of Eq. (2.40). In Chapter 2 this equation was accepted on an ad hoc basis. Actually, both (2.40) and (11.42) make good physical sense. Attempt to understand them physically by studying several problems (for example, input but no internal sources, internal sources but no input, and so on) and then using the superposition principle.

11. Show that the solutions to (11.46) and (11.49) exist provided R is a reasonably well-behaved function. (See Problems 20–24.)

12. It is remarked in Section 8 that the H-function is not completely defined by (11.56) and that there is an extraneous solution unless $\gamma=1$. Work out from first principles the analog of the H function for the semi-infinite rod. Verify by direct calculation that the same ambiguity exists in this simple case. Make a physical argument to get rid of the extraneous solution.

13. Develop invariant imbedding equations for the slab geometry in the case of energy dependence as well as angular dependence.

14. Use the general ideas of Problem 1 on the various R equations that occur in the text and in this problem section to obtain discrete imbedding equations. Compare these with the results in Chapter 1.

15. Develop a theory for transmission functions paralleling the one given in the text for reflection functions. Take note of Problem 6.

16. Obtain the analog of the fundamental equation (3.3) for the situations under current study. Use this to obtain identities of the kind (3.8) and (3.11). Describe the additional complexities encountered.

17. It is indicated in Section 9 that the invariant imbedding approach is frequently computationally superior to the direct solution of the classical Boltzmann equation. By studying the rod model try to get some idea as to why the integration of the reflection equation might be numerically more satisfactory than the solution of the u and v equations. Make an elementary study of stability, round off, and the like. Generalize to the n-state case. Note that the actual numerical computation of the R functions of this chapter always reduces the problem to an n-state model.

18. Develop time-dependent reflection equations when the various physical parameters are time-dependent. Discuss the complications introduced when k itself depends on time. Can you overcome them?

19. Show that the study of the simple delta-function input (11.57) is sufficient for the study of quite general inputs. Write out the corresponding Duhamel formulas.

20. Make a transformation similar to (11.51) to bring Eq. (11.36) into the form

$$\frac{\partial S}{\partial x}(x,\mu,\mu_1)+\left(\frac{1}{\mu}+\frac{1}{\mu_1}\right)S(x,\mu,\mu_1)=\gamma\left\{1+\frac{1}{2}\int_0^1 S(x,\mu',\mu_1)\frac{d\mu'}{\mu'}\right.$$

$$\left.+\frac{1}{2}\int_0^1 S(x,\mu,\mu'')\frac{d\mu''}{\mu''}+\frac{1}{4}\int_0^1\int_0^1 S(x,\mu',\mu_1)S(x,\mu,\mu'')\frac{d\mu'}{\mu'}\frac{d\mu''}{\mu''}\right\},\ S(0,\mu,\mu_1)=0.$$

21. Argue that the function S introduced in the previous problem is symmetric in the variables μ and μ_1; that is, $S(x,\mu,\mu_1)=S(x,\mu_1,\mu)$. Introduce the new function $\phi(x,\mu)=1+\frac{1}{2}\int_0^1(S(x,\mu,\mu_1)/\mu_1)\,d\mu_1$. Thus obtain the equation

$$\frac{\partial S}{\partial x}+\left(\frac{1}{\mu}+\frac{1}{\mu_1}\right)S(x,\mu,\mu_1)=\gamma\phi(x,\mu)\phi(x,\mu_1).$$

Integrate this to obtain a nonlinear integral equation for ϕ:

$$\phi(x,\mu)=1+\frac{\gamma}{2}\int_0^1\frac{d\mu_1}{\mu_1}\int_0^x e^{-\left(\frac{1}{\mu}+\frac{1}{\mu_1}\right)(x-t)}\phi(t,\mu)\phi(t,\mu_1)\,dt.$$

Observe that if ϕ can be found then S can be and so R can be determined.

22.* To analyze the equation for ϕ found in Problem 21 introduce a sequence of functions ϕ_n by choosing $\phi_0(x,\mu)=0$ and

$$\phi_{n+1}(x,\mu)=1+\frac{\gamma}{2}\int_0^1\frac{d\mu_1}{\mu_1}\int_0^x e^{-\left(\frac{1}{\mu}+\frac{1}{\mu_1}\right)(x-t)}\phi_n(t,\mu)\phi_n(t,\mu_1)\,dt.$$

Prove that if $\gamma\leqslant 1$ then this sequence converges for all $x\geqslant 0$ and all μ, $0<\mu\leqslant 1$. Hence obtain a proof that the corresponding ϕ, S, and R functions all exist. Show that if $\gamma>1$ then the sequence converges only for $x<x(\gamma)$. Try to get some estimates on $x(\gamma)$ as a function of γ. Give a physical interpretation for your results.

23.* Obtain some results concerning the positivity and monotonicity of ϕ_n (in x) where ϕ_n is defined above. What do these results have to say about the corresponding R function?

24.* In Problem 11 the assumption was made that R is a reasonably well-behaved function. Remark on this assumption in the light of the discoveries made in Problems 20–23.

REFERENCES

1. V. Kourganoff and I. W. Busbridge, *Basic Methods in Transfer Problems*, Dover, New York, 1963.
2. G. M. Wing, *An Introduction to Transport Theory*, Wiley, New York, 1962.
3. I. W. Busbridge, *The Mathematics of Radiative Transfer*, Tracts on Mathematics and Mathematical Physics, No. 50, Cambridge Univ. Press, London, 1960.
4. K. M. Case and P. F. Zweifel, *Linear Transport Problems*, Addison Wesley, Reading, Mass., 1967.
5. R. W. Preisendorfer, "A Mathematical Foundation for Radiative Transfer Theory," *J. Math. Mech.* **6**, 1957, 685–730.
6. J. Lehner and G. M. Wing, "On the Spectrum of an Unsymmetric Operator Arising in the Transport Theory of Neutrons," *Comm. Pure Appl. Math.* **8**, 1955, 217–234.
7. J. Lehner and G. M. Wing, "Solution of the Linearized Boltzmann Transport Equation for the Slab Geometry," *Duke Math. J.* **23**, 1956, 125–142.
8. J. Lehner, "The Spectrum of the Neutron Transport Operator for the Infinite Slab," *J. Math. Mech.* **11**, 1962, 173–182.

9. V. A. Ambarzumian, "Diffuse Reflection of Light by a Foggy Medium," *Compt. Rend. Acad. Sci. SSR.* **38**, 1943, 229.
10. S. Chandrasekhar, *Radiative Transfer*, Dover, New York, 1960.
11. M. L. Alme, *The XRAYII Code*, Technical Note WLTT-TN-70-7, Air Force Weapons Laboratory, Kirtland Air Force Base, New Mexico.
12. T. W. Mullikin, "A Complete Solution of the *X* and *Y* Equations of Chandrasekhar," *Astrophys. J.* **136**, 1962, 627–635.
13. R. Bellman, R. Kalaba, and M. C. Prestrud, *Invariant Imbedding and Radiative Transfer in Slabs of Finite Thickness*, American Elsevier, New York, 1963.
14. R. Shimizu and K. Aoki, *Applications of Invariant Imbedding to Reactor Physics*, Academic, New York, 1972.
15. J. O. Mingle, *The Invariant Imbedding Theory of Nuclear Transport*, American Elsevier, New York, 1973.
16. P. Bailey and G. M. Wing, "A Correction to Some Invariant Imbedding Equations of Transport Theory Obtained by Particle Counting," *J. Math. Anal. Appl.* **8**, 1964, 170–174.
17. R. Bellman, H. H. Kagiwada, R. Kalaba, and M. C. Prestrud, *Invariant Imbedding and Time-dependent Transport Processes*, American Elsevier, New York, 1964.

12

INTEGRAL EQUATIONS

1. INTRODUCTION

Thus far our invariant imbedding investigations have been confined mainly to equations containing derivatives. Admittedly, in the preceding chapter on transport theory and radiative transfer we were confronted with differential integro-equations, but such expressions can be thought of as limiting cases of systems of differential equations by replacing the integral term by a quadrature expression. In the present chapter we turn to the investigation of a rather wide class of integral equations of Fredholm type. Much of this work originally appeared in [1]. Not only will we discover new means of formulating such equations, using the imbedding technique, but we shall discover that this approach actually leads to new methods of computation and equally new ways of investigating their eigenvalues.

We shall begin by utilizing a very well-known equivalence between a limited class of transport problems and an equally limited class of integral equations. The reader who may have skipped the chapter devoted to transport theory need not worry about his ability to absorb the reasoning used; he need only be willing to accept the equations that arise from transport theory. By noting exactly how this equivalence comes about we shall be able to construct pseudo-transport problems which lead to a large variety of Fredholm-type integral equations. By doing the imbedding on the pseudo-transport equations and then interpreting the results in terms of the corresponding integral equations we shall have, in effect, obtained invariant imbedding equations applicable to integral equations.

Having achieved this insight we shall then turn briefly to an even wider class of Fredholm equations and obtain some results without using the simulated transport equation as an intermediary.

Relatively few numerical investigations for this kind of study have appeared in book form, although they may be found in a variety of papers. We shall therefore present a few such results as we proceed.

2. AN INTEGRAL EQUATION FOR TRANSPORT IN A SLAB

Let us turn to Eq. (11.23), and pick $Q(\mu)=\delta(\mu-\mu_0)$, $-1\leqslant \mu,\mu_0<0$. Moreover, we define

$$\phi(z)=\int_{-1}^{1} n(z,\mu)\,d\mu. \tag{12.1}$$

We rewrite (11.23a) in the form

$$\frac{\partial n}{\partial z}+\frac{n(z,\mu)}{\mu}=\frac{\gamma}{2\mu}\phi(z). \tag{12.2}$$

If we consider this as a first-order linear differential equation we obtain

$$n(z,\mu)=\frac{\gamma}{2\mu}\int^{z} e^{-(z-z')/\mu}\phi(z')\,dz'+ke^{-z/\mu}. \tag{12.3}$$

To make good use of the boundary conditions (11.23a, b) judicious choice of the integration limits is important. For $\mu>0$, we select the lower limit to be zero and obtain

$$n(z,\mu)=\frac{\gamma}{2\mu}\int_0^{z} e^{-(z-z')/\mu}\phi(z')\,dz'. \tag{12.4a}$$

For $\mu<0$, it is convenient to make the *lower* limit in (12.3) the variable z, change the sign of the integral, and pick the upper limit as x. The result is

$$n(z,\mu)=-\frac{\gamma}{2\mu}\int_z^{x} e^{-(z-z')/\mu}\phi(z')\,dz'+\delta(\mu-\mu_0)e^{-(x-z)/\mu}. \tag{12.4b}$$

Next, integrate (12.4a) on μ from zero to one and (12.4b) from minus one to zero; then add the resulting expressions:

$$\int_0^1 n(z,\mu)\,d\mu+\int_{-1}^{0} n(z,\mu)\,d\mu=\phi(z)$$

$$=\frac{\gamma}{2}\left\{\int_0^1\frac{d\mu}{\mu}\int_0^{z} e^{-(z-z')/\mu}\phi(z')\,dz'-\int_{-1}^{0}\frac{d\mu}{\mu}\int_z^{x} e^{-(z-z')/\mu}\phi(z')\,dz'\right\}$$

$$+e^{(x-z)/\mu_0}. \tag{12.5}$$

A change of order of integration is suggested. The legitimacy of this operation is of course always a matter of concern. We refer the reader to

[2] where the question is carefully studied. The result is

$$\phi(z)=\frac{\gamma}{2}\int_0^z \phi(z')\left[\int_0^1 \frac{e^{-(z-z')/\mu}}{\mu}\,d\mu\right]dz'$$

$$-\frac{\gamma}{2}\int_z^x \phi(z')\left[\int_{-1}^0 \frac{e^{-(z-z')/\mu}}{\mu}\,d\mu\right]dz' + e^{(x-z)/\mu_0}. \tag{12.6}$$

The function occuring in the first integral is well known:

$$\int_0^1 e^{-(z-z')/\mu}\frac{d\mu}{\mu}=E_1(z-z'); \tag{12.7}$$

the classical definition is

$$\int_0^1 e^{-s/w}\frac{dw}{w}=\int_1^\infty e^{-st}\frac{dt}{t}=E_1(s), \qquad \operatorname{Re}(s)>0. \tag{12.8}$$

[It is perhaps appropriate to remark that in the period before 1945 a variety of notations was used for the function defined in (12.8). (See, for example, [3].) The notation we use here is now quite standard.]

The second integral of (12.6) still seems to constitute a problem. However, we recall that there $z'>z$ and, of course, $\mu<0$. Thus,

$$-\int_{-1}^0 e^{-(z-z')/\mu}\frac{d\mu}{\mu}=\int_0^1 e^{-|z-z'|/\nu}\frac{d\nu}{\nu}= \quad E_1(|z-z'|). \tag{12.9}$$

Clearly, Eq. (12.6) now yields

$$\phi(z)=\frac{\gamma}{2}\int_0^x E_1(|z-z'|)\phi(z')\,dz'+e^{+(x-z)/\mu_0}. \tag{12.10}$$

This is an integral equation of Fredholm type. Note, however, that while the integrand is symmetric it is discontinuous along the line $z=z'$. The discontinuity is a mild one, since

$$E_1(z)\sim-\log z \tag{12.11}$$

for small but positive z.

We have thus associated with the transport equation (11.23) and the boundary conditions given at the beginning of this section the integral equation (12.10). It is a matter of some interest that it is also possible to go from the integral formulation back to the Boltzmann equation. We leave that matter to the problem section.

It is also valuable to observe that the function ϕ in (12.10) really depends on several other parameters apart from the z that is exhibited as its argument. In particular, it is a function of μ_0, which we recall lies in the interval $-1 \leqslant \mu_0 < 0$. Now let g be a reasonably arbitrary function on the same interval. (We shall be more precise later on.) Multiplying (12.10) by $g(\mu_0)$ and changing the ϕ notation to include the μ_0 dependence, we get

$$\int_{-1}^{0} \phi(z,\mu_0) g(\mu_0)\, d\mu_0$$

$$= \frac{\gamma}{2} \int_{-1}^{0} g(\mu_0)\, d\mu_0 \int_{0}^{x} E_1(|z-z'|)\phi(z',\mu_0)\, dz' + \int_{-1}^{0} e^{(x-z)/\mu_0} g(\mu_0)\, d\mu_0. \tag{12.12}$$

Finally, if we assume the legitimacy of interchange of integration order we obtain from (12.12)

$$\psi(z) = \frac{\gamma}{2} \int_{0}^{x} E_1(|z-z'|)\psi(z')\, dz' + h(z), \tag{12.13}$$

where

$$\begin{aligned} \psi(z) &= \int_{-1}^{0} \phi(z,\mu_0) g(\mu_0)\, d\mu_0, \\ h(z) &= \int_{-1}^{0} e^{(x-z)/\mu_0} g(\mu_0)\, d\mu_0. \end{aligned} \tag{12.14}$$

(In the notation used for h we have suppressed the x, considering it as constant, at least for the moment.)

Thus we see that knowledge of the solution of (12.10) automatically produces knowledge of the solution of any equation of the general form (12.13) provided the function h can be expressed by an equation like (12.14). But, knowledge concerning the solution of (12.10) *is* available by the imbedding method applied to the corresponding transport equation (12.2) as was demonstrated in Chapter 11.

It would now seem appropriate to carry through the rigorous details of the analysis we have been discussing. Instead of doing that we pause long enough to note that the same general set of ideas can be used on any equation which has the general appearance of Eq. (12.2), even though the new expression may have nothing to do with a genuine physical transport process. Therefore, instead of concentrating on the specific "realistic" problem of particles moving in a slab geometry, we shall *invent* a simulated transport equation and try to carry through the same analysis for this case that we have outlined for the slab geometry. Moreover, we shall construct our equation in such a manner that the invariant imbedding methods devised in the earlier chapters can be applied to it.

3. A PSEUDO-TRANSPORT PROBLEM AND ITS ASSOCIATED INTEGRAL EQUATION

The pseudo-transport equation we choose to examine is

$$\operatorname{sgn} s \frac{\partial n}{\partial z} + a(s)n(z,s) = k(s)\gamma(z)\int_{-\infty}^{\infty} n(z,s')\,ds', \quad \tilde{y} \leqslant y \leqslant z \leqslant x \leqslant \tilde{x}, \tag{12.15}$$

subject to the side conditions

$$n(y,s) = 0, \qquad 0 < s < \infty, \tag{12.15a}$$

$$n(x,s) = f(s), \qquad -\infty < s < 0. \tag{12.15b}$$

For the sake of precision we shall also make the following assumptions concerning the functions involved.

Assumptions A.

1. $a(s)$ and $k(s)$ are piecewise continuous even functions

   on $-\infty < s < \infty$; (12.16a)

2. $\gamma(z)$ is continuous in z for $\tilde{y} \leqslant z \leqslant \tilde{x}$; (12.16b)

3. $$\operatorname{Re}\{a(s)\} \geqslant 0, \qquad -\infty < s < \infty; \tag{12.16c}$$

4. $$\int_0^{\infty} |k(s')|\,ds' < \infty; \tag{12.16d}$$

5. $$\int_{-\infty}^{0} |f(s')|\,ds' < \infty. \tag{12.16e}$$

In Section 6 we shall find it desirable to make a somewhat more stringent set of assumptions.

Assumptions A′. These consist of Eq. (12.16a)–(12.16d) together with

5.′ f is a member of the class of functions

continuous and with compact support on

$$-\infty<s<0. \tag{12.16e′}$$

It is clear that any set of functions satisfying Assumptions A′ also satisfies Assumptions A. We shall find that for the purposes of Section 5 it is desirable to use Assumptions A′, but the generality of the final results is in no way affected by this added restriction. For the present it is appropriate to employ Assumptions A.

Unfortunately, neither conditions A nor A′ hold for the problem discussed in the previous section. There, $a(s)=2k(s)=1/|s|$ for $|s|\leqslant 1$, with $a(s)=k(s)=0$ elsewhere. Thus (12.16d) is violated and what we do in the subsequent sections is not immediately applicable to the problem of original physical interest. Although this is unfortunate, the difficulty can be overcome with some effort and ingenuity. However, should we attempt this, the presentation we are about to give would become much more abstract and sophisticated and hence not truly within the general approach of this book. We therefore shall settle for the relatively simple conditions (12.16).

We now suppose that (12.15) has a solution $n(z,s)$, which is continuously differentiable in $z, y\leqslant z\leqslant x$, and uniformly integrable in $s, -\infty<s<\infty$. (Recall that a phenomenon akin to criticality might occur for this problem which means that assumptions of this kind are truly necessary!) If we employ precisely the devices used in obtaining (12.4a) and (12.4b), we get:

$$n(z,s)=k(s)\int_y^z \gamma(z')\phi(z')e^{a(s)(z'-z)}\,dz', \qquad s>0; \tag{12.17a}$$

$$n(z,s)=k(s)\int_z^x \gamma(z')\phi(z')e^{-a(s)(z'-z)}\,dz'$$
$$+e^{a(s)(z-x)}f(s), \qquad s<0. \tag{12.17b}$$

Here we have defined

$$\phi(z)=\int_{-\infty}^{\infty} n(z,s)\,ds. \tag{12.18}$$

If we integrate Eq. (12.17a) over all positive s, and (12.17b) over all negative s, take account of the various assumptions (12.16), and generally proceed as we did in obtaining Eq. (12.13) we find eventually

$$\phi(z)=\int_y^x \gamma(z')K(|z-z'|)\phi(z')\,dz'+g(z). \tag{12.19}$$

In this equation

$$K(u)=\int_0^\infty k(s')e^{-a(s')u}\,ds', \qquad u\geqslant 0, \tag{12.20}$$

and

$$g(z)=\int_{-\infty}^0 e^{-a(s')(x-z)}f(s')\,ds'. \tag{12.21}$$

The Fredholm equation (12.19) will have a unique solution ϕ provided unity is not an eigenvalue of the operator defined by

$$T(K)=\int_y^x \gamma(z')K(|z-z'|)\cdot\,dz'. \tag{12.22}$$

At this point it seems appropriate to state a formal theorem. The result has been partially established in the foregoing; the remainder we leave to the problems (see Problem 4).

THEOREM 12.1. Let $n(z,s)$ be continuously differentiable in $z, y\leqslant z\leqslant x$, uniformly integrable in s, $-\infty<s<\infty$, and let $n(z,s)$ satisfy (12.15). Then $\phi(z)$ as defined by (12.18) satisfies (12.19). Conversely, if $\phi(z)$ is the solution to (12.19) (under the assumption that unity is not an eigenvalue of $T(K)$), then $n(z,s)$ as defined by Eqs. (12.17) is a solution of (12.15). Finally, under Assumptions A the correspondence established by this result is unique.

It will be noted that we have deliberately used a fairly general boundary condition at $z=x$, and have avoided the use of the delta-function. We further observe, however, that just as in the case of the previous section it is clear that it will suffice simply to investigate the solution to

$$\tilde{\phi}(z,s)=\int_y^x \gamma(z')K(|z-z'|)\tilde{\phi}(z',s)\,dz'+e^{-a(s)(x-z)},$$

$$-\infty<s<0, \tag{12.23}$$

and then multiply by $f(s)$ and integrate over all s, $-\infty<s<0$, in order to obtain the desired result. It is essentially the function $\tilde{\phi}(z,s)$ that we proceed to study in some detail.

4. REPRESENTATIONS FOR ϕ AND n

Henceforth we shall suppose that for no y and x in some interval $\tilde{y} \leqslant y \leqslant x \leqslant \tilde{x}$ is unity an eigenvalue of $T(K)$, and we shall always so constrain y and x. As a matter of convenience, notations like

$$n = n(z,x,y,s), \qquad \phi = \phi(z,x,y), \text{ etc.,} \tag{12.24}$$

will be used when it is desirable to emphasize dependence on variables which are usually suppressed.

The general theory of Fredholm integral operators allows us to assert the existence of a *resolvent* kernel Q,

$$Q = Q(z,z',x,y) \tag{12.25}$$

such that Eq. (12.19) is solved by

$$\phi(z,x,y) = g(z) + \int_y^x Q(z,z',x,y) g(z')\, dz'. \tag{12.26}$$

If we use the definition of g as given by (12.21) we find, after some computation,

$$\phi(z,x,y) = \int_{-\infty}^0 f(s') R(z,x,y,s')\, ds', \tag{12.27}$$

where

$$R(z,x,y,s') = e^{-a(s')(x-z)} + \int_y^x Q(z,z',x,y) e^{-a(s')(x-z')}\, dz'. \tag{12.28}$$

To obtain a similar expression for n we may use (12.26) in (12.17) and eventually obtain

$$n(z,x,y,s) = e^{a(s)(z-x)} f(s) + \int_{-\infty}^0 \hat{R}(z,x,y,s,s') f(s')\, ds', \qquad s<0; \tag{12.29a}$$

$$n(z,x,y,s) = \int_{-\infty}^0 \hat{R}(z,x,y,s,s') f(s')\, ds', \qquad s>0. \tag{12.29b}$$

Actually the explicit expressions for R and $\hat{R}$ are of no great interest. (We leave the derivation of the formula for $\hat{R}$ as Problem 6.) What is important is the following theorem.

THEOREM 12.2. The functions ϕ and n have integral representations given by Eqs. (12.27) and (12.29). Moreover, the functions R and $\hat{R}$ are continuously differentiable with respect to x, y, and z for $\tilde{y} \leqslant y \leqslant z \leqslant x \leqslant \tilde{x}$. They are bounded and piecewise continuous in s', and the only possible discontinuities in s' appear at discontinuities of k and a. The same remarks hold for s in the case of $\hat{R}$.

The proof of this result is relatively easy and is left as a problem.

5. DERIVATION OF THE PRINCIPAL RESULTS

We now seek equations for the function ϕ evaluated at $z=x$ and $z=y$. Henceforth we shall confine our investigation to classes of functions satisfying *Assumptions A'* (Section 3). This additional restriction will in no way make our final results less interesting, and it has the distinct advantage of considerably simplifying or eliminating many of the questions concerning validity of interchanges of order of integration, convergence of integrals, and so on, in the following discussions.

The information obtained in the previous section makes it clear that differentiation of (12.15) with respect to x is legitimate. (We shall frequently use the index subscript notation for partial derivatives.)

$$\operatorname{sgn} s \frac{\partial n_2}{\partial z} + a(s) n_2(z,x,y,s)$$

$$= k(s)\gamma(z)\int_{-\infty}^{\infty} n_2(z,x,y,s')\,ds', \qquad y \leqslant z \leqslant x; \tag{12.30}$$

$$n_2(y,x,y,s) = 0, \qquad s>0; \tag{12.30a}$$

$$n_2(x,x,y,s) = -n_1(x,x,y,s), \qquad s<0. \tag{12.30b}$$

This problem is completely resolved by Theorem 12.2;

$$\phi_2(z,x,y) = \int_{-\infty}^{0} R(z,x,y,s')\{-n_1(x,x,y,s')\}\,ds'$$

$$= \int_{-\infty}^{0} R(z,x,y,s')\{-a(s')n(x,x,y,s')$$

$$+ k(s')\gamma(x)\phi(x,x,y)\}\,ds'. \tag{12.31}$$

Here we have made use of (12.15). The above expression may be rewritten (in even more complicated form!) by noting that $n(x,x,y,s')$ is known from (12.15b) and that $\phi(x,x,y)$ can be replaced by the use of (12.27):

$$\phi_2(z,x,y) = -\int_{-\infty}^{0} R(z,x,y,s')a(s')f(s')\,ds'$$

$$+\gamma(x)\int_{-\infty}^{0} k(s'')R(z,x,y,s'')\,ds''\int_{-\infty}^{0} R(x,x,y,s')f(s')\,ds'. \quad (12.32)$$

Finally, we observe that the left side of (12.31) can also be written in a different way, again by the use of (12.27),

$$\phi_2(z,x,y) = \int_{-\infty}^{0} f(s')R_2(z,x,y,s')\,ds'. \quad (12.33)$$

Combining (12.32) and (12.33) yields

$$0 = \int_{-\infty}^{0} f(s')\,ds'\Big\{-R_2(z,x,y,s') - a(s')R(z,x,y,s')$$

$$+\gamma(x)R(x,x,y,s')\int_{-\infty}^{0} R(z,x,y,s'')k(s'')\,ds''\Big\}. \quad (12.34)$$

Since f is any function satisfying (12.16e′), and R is piecewise continuous in s, standard arguments lead to the equation

$$\frac{\partial R}{\partial x} = -a(s)R(z,x,y,s)$$

$$+\gamma(x)R(x,x,y,s)\int_{-\infty}^{0} R(z,x,y,s')k(s')\,ds', \quad (12.35)$$

holding at all values of s where R is continuous.

It is apparent that we have obtained an equation quite reminiscent of some that we have encountered in past investigations in that it is a partial differential integro-equation involving a "product" of R terms. However, more careful inspection reveals that one of the expressions in R involves no z argument, while all of the others do. To help overcome this difficulty we select $y = x$, which, of course, forces the choice $z = x$. From (12.27),

$$\phi(x,x,x) = \int_{-\infty}^{0} R(x,x,x,s')f(s')\,ds', \quad (12.36)$$

while from (12.18)

$$\phi(x,x,x)=\int_{-\infty}^{0} f(s')\,ds'. \tag{12.37}$$

The standard arguments now give

$$R(x,x,x,s)=1. \tag{12.38}$$

As a notational matter we set (see Problem 7 for motivation)

$$\bar{\phi}(z,x,y,s)\equiv R(z,x,y,s). \tag{12.39}$$

Our recent results now appear in the form

$$\bar{\phi}_x=\frac{\partial\bar{\phi}}{\partial x}=-a(s)\bar{\phi}(z,x,y,s)$$

$$+\gamma(x)\bar{\phi}(x,x,y,s)\int_{-\infty}^{0}\bar{\phi}(z,x,y,s')k(s')\,ds', \tag{12.40}$$

$$\bar{\phi}(x,x,x,s)=1, \qquad \tilde{y}\leqslant y\leqslant z\leqslant x\leqslant \tilde{x}.$$

Little progress seems to have been made, since Eq. (12.40) still includes one function with no z term; we have, however, obtained a side condition. It is evident that Eq. (12.40) cannot be solved by itself. Additional information is necessary. To obtain such added material we introduce a *new* problem, solving (12.15) subject to the conditions

$$n(y,x,y,s)=\tilde{f}(s), \qquad s>0, \tag{12.41a}$$

$$n(x,x,y,s)=0, \qquad s<0. \tag{12.41b}$$

(Here the function $\tilde{f}$ is required to satisfy conditions analogous to (12. 16e').) This problem can now be analyzed in just the same way as was the previous one; the details are left to the problem section. Most important is the observation that differentiation with respect to y is called for. The result is

$$\bar{\phi}_y=\frac{\partial\bar{\phi}}{\partial y}=-\gamma(y)\bar{\phi}(y,x,y,s)\int_{0}^{\infty}\underline{\phi}(z,x,y,s')k(s')\,ds'. \tag{12.42}$$

Here we observe that a new function has arisen, namely $\underline{\phi}$. In fact,

$$\underline{\phi}(z,x,y,s) = \tilde{R}(z,x,y,s), \tag{12.43}$$

where $\tilde{R}$ is the function that corresponds to R in Theorem 12.2, but which solves the problem posed with the nonzero boundary condition at y instead of at x.

A study of (12.42) reveals that the system (12.40) and (12.42) can still not be solved. But more equations are obtainable, by differentiating Eqs. (12.15) with respect to y and the problem defined by the boundary conditions (12.41) with respect to x. (See Problem 8.) The resulting set of equations is

$$\underline{\phi}_y = \frac{\partial \underline{\phi}}{\partial y} = a(s)\,\underline{\phi}(z,x,y,s)$$

$$- \gamma(y)\,\underline{\phi}(y,x,y,s)\int_0^\infty k(s')\,\underline{\phi}(z,x,y,s')\,ds', \tag{12.44}$$

$$\underline{\phi}(x,x,x,s) = 1; \qquad \tilde{y} \leqslant y \leqslant z \leqslant x \leqslant \tilde{x}.$$

$$\underline{\phi}_x = \frac{\partial \underline{\phi}}{\partial x} = \gamma(x)\,\underline{\phi}(x,x,y,s)\int_{-\infty}^0 \bar{\phi}(z,x,y,s')k(s')\,ds'. \tag{12.45}$$

The equations we have finally derived are the analogs of a very famous set obtained by Chandrasekhar in an astrophysical context [4]. Since his problems arose directly from physical considerations and involved much more symmetry than ours, his equations were substantially simpler; in fact, there were only two equations in his system rather than the four we have found it necessary to obtain. Because Chandrasekhar's equations are classically referred to as his "X and Y" equations, we choose a notation to emphasize the connection, defining

$$\bar{X}(x,y,s) = \bar{\phi}(x,x,y,s), \qquad s<0, \tag{12.46a}$$

$$\bar{Y}(x,y,s) = \bar{\phi}(y,x,y,s), \qquad s<0, \tag{12.46b}$$

$$\underline{X}(x,y,s) = \underline{\phi}(y,x,y,s), \qquad s>0, \tag{12.46c}$$

$$\underline{Y}(x,y,s) = \underline{\phi}(x,x,y,s), \qquad s>0. \tag{12.46d}$$

In terms of these X and Y functions our basic equation set now yields, with judicious choice of z:

$$\overline{Y}_1(x,y,s) = -a(s)\overline{Y}(x,y,s) + \gamma(x)\overline{X}(x,y,s)\int_{-\infty}^{0}\overline{Y}(x,y,s')k(s')\,ds', \quad (12.47a)$$

$$\overline{X}_2(x,y,s) = -\gamma(y)\overline{Y}(x,y,s)\int_{0}^{\infty}\underline{Y}(x,y,s')k(s')\,ds', \quad (12.47b)$$

$$\underline{Y}_2(x,y,s) = a(s)\,\underline{Y}(x,y,s) - \gamma(y)\,\underline{X}(x,y,s)\int_{0}^{\infty}\underline{Y}(x,y,s')k(s')\,ds', \quad (12.47c)$$

$$\underline{X}_1(x,y,s) = \gamma(x)\,\underline{Y}(x,y,s)\int_{-\infty}^{0}\overline{Y}(x,y,s')k(s')\,ds', \quad (12.47d)$$

$$\overline{X}(\xi,\xi,s) = \overline{Y}(\xi,\xi,s) = 1,$$
$$\underline{X}(\xi,\xi,s) = \underline{Y}(\xi,\xi,s) = 1, \quad (12.47e)$$
$$\tilde{y} \leqslant \xi \leqslant \tilde{x}.$$

These equations form a well-posed set; that a solution exists follows from our mode of derivation. The numerical solution of such a system remains a formidable task. Shortly we shall impose symmetry conditions which will considerably simplify the system and make calculation well within the domain of feasibility, even with computers of only moderate capacities. However, before pursuing that route it is well to stop briefly and reconsider what we have done and what we have actually accomplished.

We began by establishing an equivalence between a class of Fredholm equations and a class of pseudo-transport equations. Then, by using known results from classical integral equation theory, we rigorously applied the invariant imbedding technique to the transport-like equation and eventually emerged with the set (12.47). Of what use is this set? First, we note that once the functions $\overline{X}, \overline{Y}, \underline{X}, \underline{Y}$ are known, Eq. (12.40) is soluble for fixed z and y. Hence according to (12.39) we will have obtained the function $R(z,x,y,s)$. From the R function one can obtain, through the use of Eq. (12.27) the solution to the original integral equation (12.19).

Moreover, the calculations that must be made will readily provide the solution of (12.19) not just for a single pair of values, x and y, but for a whole set of such values (always confined to the fundamental interval $\tilde{y} \leqslant y \leqslant x \leqslant \tilde{x}$, of course). At all stages of the calculation only initial value problems are encountered. Hence the solution of a class of Fredholm integral equations is effectively reduced to the solution of a set of initial value problems.

The significance of the function $\tilde{R}$ we leave to the reader to investigate (see Problem 8).

It now seems appropriate to give a formal statement of our results.

THEOREM 12.3. The integral equation (12.19) may be completely solved for any x and y in the fundamental interval $\tilde{y} \leqslant y \leqslant x \leqslant \tilde{x}$ by solving the initial value problem (12.47), then solving the initial value problem (12.40), and finally making use of Eq. (12.27).

6. A SPECIAL CASE

In the previous section reference was made to the fact that in his original work Chandrasekhar encountered only two functions X and Y and that as a result his system of equations was less complex than ours. We shall now specialize our problem to obtain results more directly analogous to his. Let us suppose that

$$\gamma(z) = \gamma(-z), \tag{12.48}$$

$$y = -x, \tag{12.49}$$

$$f(s) = \tilde{f}(-s). \tag{12.50}$$

The symmetry considerations applied to (12.15) and the corresponding problem defined by conditions (12.41) reveal that

$$\bar{\phi}(z,x,-x,-s) = \underline{\phi}(-z,x,-x,s), \qquad s \geqslant 0. \tag{12.51}$$

Thus

$$\bar{X}(x,-x,-s) = \underline{X}(x,-x,s) \tag{12.52}$$

and

$$\bar{Y}(x,-x,-s) = \underline{Y}(x,-x,s), \qquad s \geqslant 0. \tag{12.53}$$

For notational convenience we define

$$X(x,s)=\overline{X}(x,-x,-s),$$
$$Y(x,s)=\overline{Y}(x,-x,-s), \qquad s\geqslant 0, \tag{12.54}$$

and observe that

$$X'(x,s)=\frac{\partial X}{\partial x}$$
$$=\overline{X}_1(x,-x,-s)-\overline{X}_2(x,-x,-s)$$
$$=\underline{X}_1(x,-x,s)-\overline{X}_2(x,-x,-s). \tag{12.55}$$

Now if we replace y by $(-x)$ in (12.47b) and (12.47d), s by $(-s)$ in (12.47b), recall that $k(s)$ was required to be even [Eq. (12.16a)], subtract (12.47b) from (12.47d), and remember Eqs. (12.52)–(12.54), we eventually discover

$$\frac{\partial X}{\partial x}=2\gamma(x)Y(x,s)\int_0^\infty Y(x,s')k(s')\,ds'. \tag{12.56}$$

Similarly, working with Eqs. (12.47a) and (12.47c) we compute

$$\frac{\partial Y}{\partial x}=-2a(s)Y(x,s)+2\gamma(x)X(x,s)\int_0^\infty Y(x,s')k(s')\,ds'. \tag{12.57}$$

The initial conditions are easily seen to be

$$X(0,s)=Y(0,s)=1. \tag{12.58}$$

THEOREM 12.4. When $\gamma(z)=\gamma(-z)$ then $X(x,s)=\overline{X}(x,-x,-s)$ $=\underline{X}(x,-x,s)$ and $Y(x,s)=\overline{Y}(x,-x,-s)=\underline{Y}(x,-x,s), s\geqslant 0$, satisfy (12.56)–(12.58).

In the astrophysical literature the case $\gamma=$constant is the one commonly encountered. Moreover, in that literature the basic interval is usually chosen as $(0,x)$ rather than $(-x,x)$. The classical X and Y equations therefore do not contain the factor 2 which appears in (12.56) and (12.57).

The pair of equations we have derived for X and Y is far less formidable than the four equations found in the case lacking symmetry.

Their numerical integration is relatively easy and no overwhelming computing machine power is required.

7. A NUMERICAL EXAMPLE AND SOME REMARKS ABOUT EIGENVALUES

It has been our policy in this book to pay relatively little attention to actual numerical calculations and tables of numbers. Some further discussion of the results of the previous section does seem warranted, however. Just how does one go about obtaining practical and usable information from the formulas that have been derived?

Before turning to the above question specifically, we wish to point out a by-product of the calculations we shall outline. Actually, this is a by-product of the entire theory. To avoid delving too deeply into the general theory of integral equations and their eigenvalues, let us argue largely by analogy. The interested reader will find more detailed information in the problem section (Problem 9) and the references (see [5]).

Our analysis of integral equations via invariant imbedding was inspired by the correspondence between Eqs. (11.23) and (12.10). Now (11.23) is an idealized model of an actual physical problem. We expect, on physical grounds, that if $\gamma > 1$ then for some slab thickness x the system will become critical—there will be no solution to the time-independent problem. If σ remains constant (we may as well take it to be unity) then it is reasonable to expect that as γ increases the critical thickness x_{cr} will decrease, and probably in a continuous, monotonic manner. (See Figure 12.1.) From another viewpoint, we may think of x as being fixed; there will be some (unique) multiplication constant γ_{cr}, which will just make the slab system go critical. From this perspective, γ_{cr} is the first *eigenvalue* corresponding to the given thickness x. When γ is given and x_{cr} is found then x_{cr} plays the role of the first *eigenlength* of the system. (It is again quite inappropriate here to discuss possible physical interpretations of other eigenvalues and eigenlengths.)

Because of the complete correspondence between the Boltzmann formulation and the integral formulation of the transport problem we seem to have found a way of determining the largest eigenvalue of the integral equation (we write the eigenparameter in the more-or-less standard location)

$$\lambda\psi(z) = \int_0^x E_1(|z-z'|)\psi(z')\,dz'; \tag{12.59}$$

(Since we are arguing by analogy, we ignore the fact that the E_1 kernel is not covered by our theorems). Fix $\lambda = 2/\gamma$ and vary x until (12.10) [or

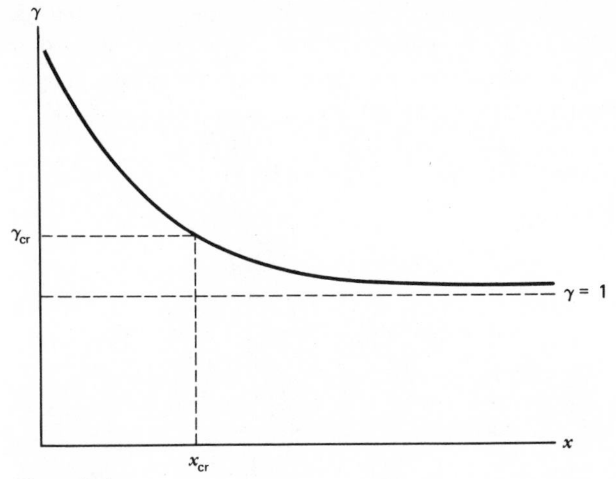

Figure 12.1.

equivalently (11.23)] ceases to have a solution. Now this is no easy task if one starts with (12.10) itself. But at this point the imbedding formulation becomes very useful, since it, after all, depends directly upon x. Hence, it seems quite reasonable to suspect that out of our results of the last few sections should emerge a method of computing eigenvalues (or eigenlengths).

Of course, the physical reasoning we have been using breaks down when the functions $k(s)$ and $a(s)$ of (12.16a) are quite general, and, as mentioned earlier, we do not wish to go into detail here as to when the method we have been describing is actually valid. Suffice it to say that in the example to be presented the calculations are justified. To keep the computations relatively easy we shall examine a problem described by Eqs. (12.56)–(12.58). We shall also assume that γ is constant so that the equation of interest is actually

$$\phi(z)=\gamma\int_{-x}^{x} K(|z-z'|)\phi(z')\,dz'+g(z). \tag{12.60}$$

(Note that the constant γ differs from that in the *physical* problem [see (12.59)] by a factor of 2, and the integration runs from $-x$ to x instead of from 0 to x.) We now integrate Eqs. (12.56)–(12.58) for given γ until X and Y fail to exist—which means, of course, that the numerical calculation will produce excessively large X and Y values. The value of x at which this occurs is the eigenlength corresponding to $\lambda=1/\gamma$, the largest eigenvalue associated with the system (12.60). The values of the X and Y functions computed up to the "blowup" point have the significance described in Sections 6. Thus the eigenlength is in truth a by-product of our calculations. By varying γ and finding the corresponding eigenlength x a curve similar to Figure 12.1 can be obtained.

Now let us turn briefly to a specific example and some sketchy remarks concerning the actual numerics involved. (For further details, see [5].) We choose

$$a(s)=|s|, \qquad -\infty<s<\infty; \tag{12.61a}$$

$$k(s)=\begin{cases} e^{-5|s|} & |s|\leqslant 1, \\ 0 & |s|>1. \end{cases} \tag{12.61b}$$

Thus,

$$K(|z-z'|)=\frac{1-e^{-(5+|z-z'|)}}{5+|z-z'|}. \tag{12.62}$$

Obviously, the integrals in (12.56) and (12.57) must be replaced by some sort of a quadrature. This is a relatively easy task in the case at hand since the nature of the function k makes the interval of integration finite. (The more general problem may still be handled, of course, if some care is taken to choose a good quadrature formula.) There is a strong tendency in the literature to choose Gaussian quadrature as the method of integral approximation. This undoubtedly stems from the origins of the problem; in astrophysics the integral is genuinely an integral over the cosine of an angle. In the type of general problem currently under consideration there is no such physical compulsion. The quadrature choice should be dictated more by the actual numerical and calculational considerations than by any physical motivation. In any event the approximate equations that replace (12.56) and (12.57) have the form

$$\frac{\partial X_i}{\partial x}=2\gamma Y_i(x)\sum_{p=1}^{N} Y_p(x)k_p w_p, \tag{12.63}$$

and

$$\frac{\partial Y_i}{\partial x}=-2a_i Y_i(x)+2\gamma X_i(x)\sum_{p=1}^{N} Y_p(x)k_p w_p, \tag{12.64}$$

where we have written $X_i(x)$ for $X(x,s_i)$, a_i for $a(s_i)$, and so on, the s_i being the quadrature points, and the w_i the corresponding weights.

Clearly, one must integrate the system (12.63), (12.64) subject to the initial conditions

$$X_i(0)=Y_i(0)=1, \qquad i=1,2,\ldots,N. \tag{12.65}$$

Again, the integration method is quite arbitrary, and is best dictated by the "shape" of the functions involved, the computing equipment available, and the like. In our example, with k and a given by (12.61) Simpson's rule was used to do the quadrature (thus determining the weights w_i) and the integration of the system was accomplished with an Adams–Bashforth method, using Runge–Kutta as a starter. Since near an eigenlength one expects the X and Y functions to get large (and it seems on an experimental basis that their growth near an eigenlength is spectacularly rapid) the integration interval in the Adams–Bashforth method is best varied as this eigenlength is approached.

Table 1 gives the value of X_j and Y_j as functions of x, where j is that index corresponding to $s_j = 0.5$. The value of γ chosen was 1.0. In Simpson's rule 21 quadrature points were taken ($N = 21$) and the computing interval was 2^{-4}. Other choices of N and the computing interval were also tried. The method is not very sensitive to such changes. As a further check the original integral equation (12.60) with $g(z)$ properly chosen (see Problem 10) was solved by a very classical iteration method. This, of course, produces the value of $\phi(z)$ at all intermediate points $-x \leqslant z \leqslant x$ used in the iteration scheme. We have indicated by $\hat{X}$ and $\hat{Y}$ the values obtained for $\phi(x)$ and $\phi(-x)$, using this iteration technique.

TABLE 1

x	X	Y	$\hat{X}$	$\hat{Y}$
0.125	1.047957	0.9304089	1.047956	0.9304077
0.250	1.092981	0.8714578	1.092978	0.8714551
0.500	1.176765	0.7811713	1.176758	0.7811648
0.750	1.255813	0.7222455	1.255802	0.7222342
1.000	1.334050	0.6901822	1.334039	0.6901644

Eigenlengths have also been computed, using smaller and smaller computing intervals as "blow-up" was approached. Rather arbitrarily, "blow-up" was defined as X_j or Y_j exceeding 10^{25}. Additional numerical experiments indicate that choosing the value of $X(x,0.5)$ or $Y(x,0.5)$ as a criterion for determining x_{cr} is a reasonable one; all components X_i and Y_i seem to get large simultaneously. There is much need for further numerical studies, both analytical and experimental, to be made on this type of problem. Table 2 lists some results. The eigenlengths thus obtained agree quite well with those calculated or estimated by more classical methods. The digital computer used was one of relatively small capacity, and much better values for x_{cr} would be obtainable on more modern machines.

Moreover, as this is being written various new devices for handling Eqs. (12.63) and (12.64) are being investigated. These promise greater accuracy and much more rapid calculations.

TABLE 2

$1/\gamma$	x_{cr}
0.025	0.06836
0.050	0.13281
0.100	0.26465
0.150	0.40137
0.200	0.54199
0.300	0.83691
0.400	1.15137

8. FURTHER REMARKS ABOUT THE FOREGOING

The work we have carried out in this chapter raises numerous questions. One might ask, for instance, if it necessary to have four equations for the functions $\overline{X}, \overline{Y}, \underline{X}, \underline{Y}$. Inasmuch as a single equation was obtainable for the reflection function in earlier chapters (and $\overline{X}$ is analogous to such a function) it is reasonable to expect that a single equation should be obtainable for the function $\overline{X}$. Such an equation has been found [1, 6]. We shall merely state it and leave the details of the derivation to the reader.

$$\overline{X}(x,y,-s)=1+\int_0^\infty k(s')\,ds'\int_y^x \gamma(x')\overline{X}(x',y,-s)$$
$$\times\ \overline{X}(x',y,-s')\exp[-\{a(s)+a(s')\}(x-x')]\,dx', \qquad s\geqslant 0. \quad (12.66)$$

We also might well ask if the whole eigenvalue question cannot be further clarified. And should not the basic functional equations discussed in Chapter 3, properly modified to take into account the much more complicated situation with which we are now dealing, be useful in the kind of investigations we have making? After all, they contain precisely the quantities of interest, the reflection and transmission functions for both right and left inputs—the analogs of the $\overline{X}, \overline{Y}, \underline{X}, \underline{Y}$ functions. R. C. Allen, Jr., has provided a rather complete investigation of many of these questions [5]. To go deeply into his methods would again take us far afield. A few of his ideas are left to the problems (see Problem 11).

The authors are not aware of any actual numerical calculations having been made with the full set of Eqs. (12.47). (For recent results, see [10]).

However, computing machines which will handle such complicated systems are now available, and some numerical experiments would be interesting and valuable. The structure of these equations, and of the simpler set (12.56)–(12.57) for that matter, is such that the imbedding approach seems of genuine value only if one is interested in a *family* of integral equations of the form (12.19) with x and y as parameters. If only one or two pairs of (x,y) values need be studied the classical iteration method is likely to be far superior to the imbedding approach, unless one happens to be in the vicinity of an eigenlength. Even in such an instance, other numerical devices are available. Of course, this same phenomenon is true in most of the problems we have examined in this book. The imbedding method seems to be most appropriate when one is studying a whole family of problems depending upon a parameter which characterizes the "size" of the system. However, as already mentioned, certain astrophysical problems have only been resolved numerically by the imbedding method, largely because the method frequently possesses a numerical stability far greater than many of the more common schemes.

9. A COMPLETELY DIFFERENT APPROACH

In this final section we shall sketch a quite different attack. No connection with the pseudo-transport equation will be made; indeed, we can allow the kernel involved in the integral equation to be much more general than in the case we have been studying. By now the reader will note that the method used is really one of the many versions of invariant imbedding. In this section we shall make no attempt at rigor, leaving that matter, together with some applications of the result, as fit material for the problems.

Let us consider the linear integral equation

$$\phi(z)=g(z)+\int_y^x K(z,z')\phi(z')\,dz'. \tag{12.67}$$

We shall make no assumptions on the kernel K nor the function g, save that their behavior is "good" enough to allow the formalities that follow. In an earlier notation [see Eq. (12.26)] Eq. (12.67) is solved by

$$\phi(z)=\phi(z,x,y)=g(z)+\int_y^x Q(z,z',x,y)g(z')\,dz'. \tag{12.68}$$

In what follows we shall emphasize the dependence of the various quantities on the lower limit of integration, y, and shall simply delete the x dependence from the functions that arise.

If we differentiate (12.67) formally with respect to y and use the notation

$\phi=\phi(z,y)$, we obtain

$$\phi_2(z,y)=-K(z,y)\phi(y,y)+\int_y^x K(z,z')\phi_2(z',y)\,dz'. \qquad (12.69)$$

Equation (12.68) now yields

$$\phi_2(z,y)=-K(z,y)\phi(y,y)-\int_y^x Q(z,z',y)K(z',y)\phi(y,y)\,dz'. \qquad (12.70)$$

But

$$\phi(y,y)=g(y)+\int_y^x Q(y,z',y)g(z')\,dz', \qquad (12.71)$$

which allows (12.70) to be rewritten as

$$\phi_2(z,y)=-\left[K(z,y)+\int_y^x Q(z,z',y)K(z',y)\,dz'\right]$$

$$\times\left[g(y)+\int_y^x Q(y,t,y)g(t)\,dt\right]. \qquad (12.72)$$

Another expression for $\phi_2(z,y)$ may be obtained by differentiating (12.68) with respect to y,

$$\phi_2(z,y)=-Q(z,y,y)g(y)+\int_y^x Q_3(z,z',y)g(z')\,dz'. \qquad (12.73)$$

Equations (12.72) and (12.73) yield

$$-g(y)\left[K(z,y)+\int_y^x Q(z,z',y)K(z',y)\,dz'-Q(z,y,y)\right]$$

$$=K(z,y)\int_y^x Q(y,t,y)g(t)\,dt+\int_y^x\int_y^x Q(z,z',y)Q(y,t,y)K(z',y)g(t)\,dz'\,dt$$

$$+\int_y^x Q_3(z,t,y)g(t)\,dt. \qquad (12.74)$$

We may now employ an argument reminiscent of a standard one used in the calculus of variations. Since $g(z)$ is an arbitrary function we choose it so that $g(y)=0$. Then (12.74) becomes

$$\int_y^x \left[K(z,y)Q(y,t,y) + \int_y^x Q(z,z',y)Q(y,t,y)K(z',y)\,dz' + Q_3(z,t,y) \right] g(t)\,dt = 0. \tag{12.75}$$

Because g has been specified only at $z=y$ we may choose it as we like elsewhere. Assuming continuity of the functions K and Q (see Problem 12) we conclude

$$Q(y,t,y)\left[K(z,y) + \int_y^x Q(z,z',y)K(z',y)\,dz' \right] = -Q_3(z,t,y). \tag{12.76}$$

Returning to (12.74) we now discover that

$$Q(z,y,y) = K(z,y) + \int_y^x Q(z,z',y)K(z',y)\,dz'. \tag{12.77}$$

This is just the classical resolvent equation (see, e.g., [7]).

Substituting (12.77) in (12.76) gives a Riccati-like equation:

$$Q_3(z,t,y) = \frac{\partial Q}{\partial y}(z,t,y) = -Q(y,t,y)Q(z,y,y). \tag{12.78}$$

Equation (12.78) is relatively new and is frequently referred to as the Bellman–Krein equation [8, 9]. It is the invariant imbedding equation we have been seeking.

We shall leave the matter of application of this new equation largely to the problems. It is interesting to note its form when the original kernel K is symmetric. Then the resolvent Q is also symmetric and (12.78) becomes

$$\frac{\partial Q}{\partial y}(z,t,y) = -Q(y,z,y)Q(y,t,y). \tag{12.79}$$

Thus if $Q(y,\xi,y)$ is known for all ξ, $y \leqslant \xi \leqslant x$, a simple integration will yield the complete resolvent.

10. SUMMARY

In this final chapter we have applied the method of invariant imbedding to the solution of a wide class of integral equations. Most emphasis has been placed on those integral equations which in some sense "mimic" the classical equation of radiative transfer. This class is by no means small nor unimportant, and many of the equations encountered in modern applications far removed from the field of transport theory are covered by the method we have devised. As this is being written, further numerical experimentation is being done and it seems hopeful that the full scale problem without the special structure required in Section 7 is amenable to computation with the computing machines now available.

We have also indicated a completely different approach, quite formal and apparently devoid of physical origins. Although the mathematics involved can be made rigorous, the physical meaning of the manipulations eludes us. Greater understanding might produce even more general methods of this nature.

In conclusion, we hasten to reiterate that our effort in writing this book has not been to record all that has been accomplished with the method of invariant imbedding. Rather, we have tried to present to the reader the basic ideas involved in the theory and then to show some applications. We apologize for oversights and omissions. However, we do feel that anyone who has absorbed a reasonable amount of the material we have presented is now in a position to consult the literature and pursue the subject further with reasonable ease and understanding.

PROBLEMS

1. Let ϕ be a solution to Eq. (12.10). Define a function n by means of Eqs. (12.4a) and (12.4b). Prove that n satisfies (12.2). Discuss the legitimacy of the operations that must be carried out for this verification.

2. Under what conditions on the various functions involved are the manipulations used to obtain (12.12) valid?

3.* It is mentioned in Section 3 that the theory developed in this chapter does not hold for Eq. (12.10), the very equation that led to this theory. Try to modify the reasoning and assumptions that have been used in such a way as to cover this classical case.

4. Obtain (12.19) and then prove Theorem 12.1.

5. The assumption is made at the beginning of Section 4 that the operator $T(K)$ does not have eigenvalue unity for any x and y such that $\tilde{y} \leqslant y \leqslant x \leqslant \tilde{x}$. In the light of previous restrictions of this kind it seems reasonable that this assumption is connected with the "criticality" of the pseudo-transport model that has been introduced. Discuss this, referring to Chapter 11 and earlier work where appropriate.

6. Find an equation for $\hat{R}$; then prove Theorem 12.2.

7. In the literature of astrophysics it is customary to consider the slab of transport material as being horizontal, with radiation impinging on the top $(z=x)$ or the botton $(z=y)$. Show that in our pseudo-transport model $\bar{\phi}(z,x,y,s)$ is the total amount of "radiation" at z due to an input at the "top" in "direction" s. Similarly interpret $\underline{\phi}$.

8. Obtain Eqs. (12.42) and (12.43), discussing the function $\tilde{R}$ in some detail. Then derive (12.44) and (12.45).

9.* Consider the arguments concerning eigenvalues, eigenlengths, and the like, as made in Section 7 from a rigorous point of view. Show that γ is indeed a decreasing function of x_{cr} under reasonable assumptions. Is it also continuous? Since there is no need here to rely on physical analogy it should also be possible to discuss higher eigenvalues.

10. Outline the classical iteration method for solution of integral equations (see Section 7). Make a comparison between its possible usefulness and that of the method given in this Chapter. If possible, program an example different from that of the text using both methods and make some comparative calculations.

11. Obtain the analogs of Eqs. (3.3). In the current case the simple reciprocals found in (3.5) (or inverses in the corresponding matrix case) become much more formidable operators. Discuss their properties and the resulting computational difficulties. (See [10].)

12. Obtain the Bellman–Krein formula rigorously, imposing reasonable conditions on the functions involved.

13.* Given the integral equation

$$u(t)=g(t)+\lambda\int_0^{\infty}e^{-a|t-t'|}u(t')\,dt', \qquad a>0,$$

where λ is not an eigenvalue. Consider the truncated equation

$$v(t)=g(t)+\lambda\int_0^{t}e^{-a|t-t'|}v(t')\,dt'.$$

If $g(t)\to 0$ as $t\to\infty$ does $|u(t)-v(t)|\to 0$ as $t\to\infty$?

14. Consider the inhomogeneous Sturm–Liouville system

$$L_\lambda y = Ly+\lambda y=(p(t)y'(t))'+(-q(t)+\lambda)y(t)=-f(t),$$

$$a<t<1,\ y(a)=0, \qquad y(1)+\beta y'(1)=0.$$

The solution may be written in terms of the Green's function for the operator L_λ provided λ is not an eigenvalue of $(-L)$:

$$y(t)=\int_a^1 G(t,s,\lambda,a)f(s)\,ds.$$

Prove that

$$\frac{\partial G}{\partial a}(t,s,\lambda,a)=p(a)\frac{\partial G}{\partial t}(a,s,\lambda,a)\frac{\partial G}{\partial s}(t,a,\lambda,a).$$

15.* The function G in the preceding problem is known to be meromorphic with simple poles at the eigenvalues of the differential operator $(-L)$. If $u_n(t)$ are the normalized eigenfunctions and λ_n are the eigenvalues of that operator then, in fact,

$$G(t,s,\lambda,a)=\sum_{n=1}^{\infty}\frac{u_n(t)u_n(s)}{(\lambda-\lambda_n)}.$$

The functions u_n and the eigenvalues λ_n are, of course, dependent upon a although that dependence has not been exhibited. Show, however, that

$$\frac{d\lambda_n}{da}=p(a)u_n'(a)^2,$$

$$\frac{\partial u_n}{\partial a}(t)=p(a)u_n'(a)\sum_{m\neq n}^{\infty}\frac{u_m(t)u_m'(a)}{\lambda_n-\lambda_m}.$$

(Here $u_n'(a)=u_n'(t)|_{t=a}$.) See:

R. Bellman and S. Lehman, "Functional Equations in the Theory of Dynamic Programming-X: Resolvents, Characteristic Functions and Values," *Duke Math. J.* **27**, 1960, 55–70;

R. Bellman and S. Lehman, "Functional Equations in the Theory of Dynamic Programming-VIII: The Variation of the Green's Function for the One-Dimensional Case," *Proc. Nat. Acad. Sci. USA* **43**, 1957, 339–341;

R. Bellman and S. Lehman "Functional Equations in the Theory of Dynamic Programming-IX: Variational Analysis, Analytic Continuation, and Imbedding of Operators," *Proc. Nat. Acad. Sci. USA* **44**, 1958, 905–907.

16. Consider the homogeneous linear integral equation

$$u(t)=\lambda\int_0^1 k(t,s')u(s')\,ds'.$$

Let $Q(t,s,\lambda)$ be the associated resolvent kernel. It satisfies the resolvent equations

$$Q(t,s,\lambda)=k(t,s)+\lambda\int_0^1 k(t,s')Q(s',s,\lambda)\,ds'$$

and

$$Q(t,s,\lambda)=k(t,s)+\lambda\int_0^1 k(s',t)Q(t,s',\lambda)\,ds'.$$

Prove that

$$\frac{\partial Q}{\partial \lambda}(t,s,\lambda)=\int_0^1 Q(t,s',\lambda)Q(s',s,\lambda)\,ds'.$$

(This result is due to Kalaba and Casti. For a demonstration, application and further references, see M. A. Golberg, "An Initial Value Method for the Computation of Characteristic Values and Functions of an Integral Operator," *J. Math. Anal. Appl.* **40**, 1972, 625–633.)

REFERENCES

1. G. M. Wing, "On Certain Fredholm Integral Equations Reducible to Initial Value Problems," *SIAM Rev.* **4**, 1967, 655–670.
2. J. Lehner and G. M. Wing, "On the Spectrum of an Unsymmetric Operator Arising in the Transport Theory of Neutrons," *Comm. Pure Appl. Math.* **8**, 1955, 217–234.
3. E. Jahnke and F. Emde, *Tables of Functions with Formulae and Curves*, 4th ed., Dover, New York, 1945.
4. S. Chandrasekhar, *Radiative Transfer*, Dover, New York, 1960.
5. R. C. Allen, Jr., "Functional Relationships for Fredholm Integral Equations Arising from Pseudo-transport Problems," *J. Math. Anal. Appl.* **30**, 1970, 48–78.
6. S. Ueno, "Diffuse Reflection and Transmission in a Finite Inhomogeneous Atmosphere," *Astrophys. J.* **132**, 1960, 729–735.
7. J. A. Cochran, *Analysis of Linear Integral Equations*, McGraw Hill, New York, 1972.
8. R. Bellman, "Functional Equations in the Theory of Dynamic Programming. VII: A Partial Differential Equation for the Fredholm Resolvent," *Proc. Amer. Math. Soc.* **8**, 1957, 435–440.
9. M. G. Krein, "On a New Method for Solving Integral Equations of the First and Second Kinds," *Dokl. Akad. Nauk, SSSR* **100**, 1955, 413–416.
10. G. P. Boicourt, "A Method of Moments Applied to the Invariant Imbedding Solution of a Certain Class of Fredholm Integral Equations," Los Alamos Scientific Laboratory Report LA-5356T, 1973.

AUTHOR INDEX

Allen, R.C., Jr., 49, 53, 168, 184, 185, 245
Alme, M.L., 218
Ambarzumian, V.A., 1, 21, 31, 186, 196, 197, 201, 202, 206, 218
Angel, E., 87
Aoki, K., 218
Atkinson, F.V., 107

Bailey, P.B., 146, 218
Bateman, H., 132
Bellman, R., 20, 37, 38, 52, 53, 66, 85, 86, 87, 107, 132, 149, 157, 167, 184, 215, 218, 241, 243, 244, 245
Beyer, W.A., 168
Boicourt, G.P., 245
Boltzmann, L., 4, 16, 32, 130, 187, 188, 189, 209, 210, 211, 215, 216
Bremmer, H., 88, 100, 102, 105, 106, 107
Brown, T.A., 38, 53
Burger, H.C., 51, 131
Burgmeier, J.W., 85
Busbridge, I.W., 208, 217

Calogero, F.A., 168
Carleman, S., 20
Case, K.M., 217
Casti, J.L., 245
Chandrasekhar, S., 1, 16, 17, 21, 31, 186, 202, 208, 218, 230, 232, 245
Cherry, I., 37
Cochran, J.A., 245
Coddington, E., 66
Collatz, L., 143, 146
Cooke, K., 37, 66

Denman, E.D., 53
Dirac, G., 111, 129, 191
Drobnies, I., 85
Duhamel, R., 124, 214, 216

Edlund, M.C., 132
Emde, F., 245
Euler, L., 87

Floquet, C., 182, 184
Fredholm, I., 81, 134, 145, 219, 221, 225, 226, 231, 232, 245

Gauss, C.F., 236
Glasstone, S., 132
Goldberg, M.A., 245
Green, H., 243, 244
Gronwall, T.H., 149, 157, 167
Gudonov, S.K., 146

Hagin, F.G., 166, 168
Harris, T., 87
Hermite, C., 166

Ince, E.L., 146

Jahnke, E., 245

Kagiwada, H.H., 218
Kalaba, R., 38, 66, 107, 132, 215, 218, 245
Kourganoff, V., 217
Krein, M.G., 241, 243, 245
Kronecker, L., 12, 76
Kutta, R., 142, 164, 178, 237

Lagrange, R., 110, 189
Laguerre, C., 19
Laplace, P.S., 7, 108, 113, 114, 115, 116, 120, 122, 126, 127, 128, 129, 130, 131, 210, 212
Latzko, H., 146
Lax, P.D., 131
Lehman, S., 244
Lehner, J., 217, 245
Levinson, N., 66

Mathieu, L., 185
Meyer, G.H.F., 38
Mingle, J.O., 38, 218
Mullikin, T., 208, 218
Mundorff, P., 185

Neumann, C., 85, 86, 87, 106
Newton, I., 127

Poincaré, H., 87
Polya, G., 50
Preisendorfer, R.W., 53, 217
Prestrud, M.C., 216, 218
Prüfer, O., 140, 141, 142, 145

Redheffer, R.M., 1, 21, 53
Reid, W.T., 53
Ricatti, L., 16, 17, 19, 20, 32, 34, 35, 36, 98, 136, 144, 145, 197, 202, 205, 241
Richardson, J.M., 85
Riemann, B., 189
Riesz, F., 146
Runge, K., 142, 164, 178, 237
Ryabenki, V.S., 146
Rybicki, G., 32, 38

Schmidt, H.W., 1, 8, 17, 20
Schrödinger, L., 147, 168
Scott, M.R., 146
Shampine, L.F., 146
Shimizu, R., 218
Simpson, R., 237
Stieltjes, J., 191, 215
Stokes, G.C., 1, 20, 88, 91, 93, 97, 107
Sz-Nagy, B., 146

Titchmarsh, E.C., 168

Ueno, S., 245
Usher, P.D., 38

Van de Hulst, H.C., 44, 53, 180, 185
Vasudevan, R., 86

Weinberg, A.M., 132
Wigner, E.P., 132
Wing, G.M., 21, 37, 38, 49, 66, 132, 146, 168, 184, 185, 215, 217, 218, 245
WKB (Brillouin, Kramers, Wentzel), 88

Zweifel, P.F., 217

SUBJECT INDEX

Approximation, 88, 98-100, 215, 236

Beta-ray, 1, 186
Boundary value problem, 4, 5, 15, 38, 50, 53-55, 60, 64, 87, 125, 138, 145, 169, 170, 182, 184

Collision, 2, 5-11, 13, 14, 22, 38, 54, 55, 60, 110, 112, 134, 187-190, 196, 215
Computation, 83, 84, 192, 193, 209, 210, 219, 235, 242, 243, 245
Conservative, 54, 55, 59, 62, 65
Critical, 5, 17, 54, 65, 134, 142, 146

Difference equations, 21, 37, 69, 72, 73, 77, 79, 171, 172, 174, 176-178, 182, 183
Differential equations, 5, 6, 8, 12, 14, 16, 18, 29, 30, 32, 35, 36, 44-50, 52, 53, 56, 57, 59-61, 66, 69, 73, 87, 98, 99, 111, 131, 145, 146, 151-154, 158, 161, 166, 168, 170, 171, 177, 178, 182, 219, 220
Discrete, 67, 70, 83, 216
Dissipation, 55, 60-62, 64, 66
Duality, 186

Eigenlength, 133, 134, 136, 139, 140, 142, 145, 183, 184, 234, 235-237, 239, 243
Eigenvalues, 133, 138, 139, 142, 144, 145-147, 154, 155, 158, 160-166, 168, 183, 184, 219, 225, 226, 234, 235, 238, 242-244

Fission, 106, 142, 187, 190
Functional equation, 1, 21, 35, 39, 40, 43-45, 48-50, 64, 73, 87, 107, 126, 238, 244, 245

Gamma rays, 186

Imbedding, 1, 8-10, 16-18, 24, 26, 32, 35, 37, 39, 49, 51, 76, 83, 87, 89, 91, 95, 96, 103, 105, 109, 114, 116, 124-127, 133, 141, 145, 147, 151, 154, 157, 166, 179, 186, 193, 206, 208-210, 214, 216, 219, 222, 223, 235, 239, 244
Integral equations, 52, 53, 59, 87, 106, 149, 219, 220, 243
Invariant imbedding, 1, 4, 6, 9, 13, 16, 18, 19, 22, 23, 25, 28, 31, 32, 35-38, 47, 49, 51-54, 64, 66, 67, 70, 71, 74, 77, 79, 81, 82, 84, 87, 88, 105, 107, 108, 111, 116, 117, 123-125, 129, 133, 141, 144-147, 149, 154, 166, 168, 169, 171, 182, 184, 185, 188, 193, 196, 197, 202, 205, 208-210, 214, 216, 218, 219, 231, 234, 239, 241, 242, 245
Inversion, 109, 127

Laplace transform, 108, 113, 114, 116, 119, 122, 126, 127,

129-131, 214
Light reflection, 1
Linear, 5, 6, 16, 20, 23, 27, 28-30, 39, 52, 53, 56, 64, 69, 81, 85-87, 89, 98, 106, 124, 134, 145, 152, 160, 205, 206, 217, 220, 239, 244, 245

Nonconservative, 55, 60, 63, 65, 66
Numerical, 83, 138, 140, 162, 164-166, 169, 178, 182, 183, 185, 191, 206, 210, 214, 216, 219, 231, 234-237, 239, 242
Numerical method, 127, 169, 179

Particle, 2-11, 13, 14, 16, 18, 22, 24, 33, 34, 40, 41, 43, 44, 49, 54, 57, 58, 60-62, 64, 65, 67, 68, 70, 73-76, 78, 80-82, 84, 109-112, 114, 116-119, 129, 134, 142, 146, 169, 186-193, 196-198, 208, 210, 211, 215, 218, 223
Particle counting, 4, 13, 23-27, 42, 44, 45, 48, 61, 64, 88, 108, 109, 113, 116, 117, 129-131, 197, 202, 209, 215
Periodic coefficients, 169, 183
Periodic media, 169, 170, 185
Phase shift, 147, 149, 154-156, 158, 164-166
Photons, 186, 187
Principles of invariance, 1

Quadrature, 208, 215, 219, 236, 237
Quantum mechanical, 147, 154

Radiative transfer, 1, 21, 91, 186-188, 216-219, 242, 245
Random walk, 67, 68, 70, 73, 74, 76, 81-84, 86, 108
Reactor, 186
Reflection, 5, 6, 13, 15, 17, 21, 25, 27, 29, 32, 34-36, 39-42, 44, 56, 90, 91, 93-97, 99-103, 105, 106, 118, 119, 120, 124, 127, 129, 130, 131, 198, 202, 205, 206, 209, 213, 215, 216, 238, 245

Schrödinger, 147, 168
Sturn-Liouville, 133, 138, 146

Time-dependent, 77, 109, 111, 113, 114, 116, 118-121, 129-132, 192, 210, 211, 215, 216, 218
Time-dependent, 2, 4, 11, 23, 40, 76, 109, 111, 113, 114, 131, 191, 193, 196, 197, 202, 215, 234
Transmission, 1, 5, 6, 13-15, 17, 21, 25, 27, 29, 35, 36, 39, 40, 42, 48, 53, 88, 90, 91, 94, 95, 100, 102, 103, 105, 106, 118, 120, 124, 127, 130, 131, 205, 215, 216, 238, 245
Transport, 16, 23, 25, 34, 38, 40, 43, 44, 49, 53, 54, 60, 61, 67, 68, 74, 88, 103, 108-110, 114, 116, 120, 121, 129, 130-132, 134, 142, 146, 169, 180, 185, 186, 188, 190-194, 196, 206, 208, 209-211, 214, 215, 217-223, 231, 234
Transport theory, 21, 37, 49, 62, 90, 129, 186, 187, 190-193, 209, 242-245

Variable coefficients, 6,54,64

Wave, 22, 88-92, 94, 95, 98, 100-103, 105, 106, 120, 121, 128, 131, 187
Wavelet counting, 88, 91
Wave propagation, 88, 103, 105, 107, 169
Wave propagation and diffusion, 109

X-ray, 186